MW00835267

3D Industrial Printing with Polymers

Scrivener Publishing
100 Cummings Center, Suite 541J
Beverly, MA 01915-6106

Publishers at Scrivener
Martin Scrivener (martin@scrivenerpublishing.com)
Phillip Carmical (pcarmical@scrivenerpublishing.com)

3D Industrial Printing with Polymers

Johannes Karl Fink

WILEY

This edition first published 2019 by John Wiley & Sons, Inc., 111 River Street, Hoboken, NJ 07030, USA
and Scrivener Publishing LLC, 100 Cummings Center, Suite 541J, Beverly, MA 01915, USA

For more information about Scrivener publications please visit www.scrivenerpublishing.com.

Wiley Global Headquarters
111 River Street, Hoboken, NJ 07030, USA

For details of our global editorial offices, customer services, and more information about Wiley products visit us at www.wiley.com.

Library of Congress Cataloging-in-Publication Data
ISBN 978-1-119-55526-1

Cover images: Pixabay.Com
Cover design by: Russell Richardson

Set in size of 11pt and Minion Pro by Exeter Premedia Services Private Ltd., Chennai, India

10 9 8 7 6 5 4 3 2 1

Contents

Preface

The scientific literature with respect to 3D printing is collected in this monograph. The text focuses on the basic issues and also the literature of the past decade. The book provides a broad overview of 3D printing procedures and the materials used therein. In particular, the methods of 3D printing are discussed initially. Then, the polymers and composites used for 3D printing are detailed.

Furthermore, the fields of uses are discussed. The main fields are electric and magnetic uses, medical applications, and pharmaceutical applications. Electric and magnetic uses also include electronic materials, actuators, piezoelectricmaterials, antennas, batteries and fuel cells.

A special chapter deals with aircraft and automotive uses for 3D printing, such a in manufacturing of aircraft parts, aircraft cabins, and others. In the field of cars, 3D printing is gaining importance for automotive parts (brake components, drives), for the fabrication of automotive repair systems, and even 3D printed vehicles.

Medical applications include organ manufacturing, bone repair materials, drug-eluting coronary stents, and dental applications. Finally, pharmaceutical applications include composite tablets, transdermal drug delivery, and patient-specific liquid capsules.

How to Use This Book

Utmost care has been taken to present reliable data. Because of the vast variety of material presented here, however, the text cannot be complete in all aspects, and it is recommended that the reader study the original literature for more complete information.

Index

There are three indices: an index of acronyms, an index of chemicals, and a general index. In the index of chemicals, compounds that occur extensively are not included at every occurrence, but rather when they appear in an important context. When a compound is found in a figure, the entry is marked in boldface letters in the chemical index.

Acknowledgements

I am indebted to our university librarians, Dr. Christian Hasenhüttl, Dr. Johann Delanoy, Franz Jurek, Margit Keshmiri, Dolores Knabl Steinhäufl, Friedrich Scheer, Christian Slamenik, Renate Tschabuschnig, and Elisabeth Groß for their support in literature acquisition. In addition, many thanks to the head of my department, ProfessorWolfgang Kern, for his interest and permission to prepare this text.

I also want to express my gratitude to all the scientists who have carefully published their results concerning the topics dealt with herein. This book could not have been otherwise compiled.

Last, but not least, I want to thank the publisher, Martin Scrivener, for his abiding interest and help in the preparation of the text. In addition, my thanks go to Jean Markovic, who made the final copyedit with utmost care.

Johannes Fink
Leoben, 16th October 2018

1

Methods of 3D Printing

The issues of 3D printing have been summarized (1). There are recent monographs on 3D printing (2–19). Also, the technical and economic development and the advances of 3D printing have been summarized, as well as detailed examples of applications (20).

Three-dimensional printing can fabricate 3D structures by inkjet printing with a liquid binder solution printed onto a powder bed (21,22)

A wide range of materials could be utilized in printing since most biomaterials exist in either a solid or liquid state. The process begins by spreading a layer of fine powder material evenly across the piston. The X-Y positioning system and the printhead are synchronized to print the desired 2D pattern by selective deposition of binder droplets onto the powder layer (23). The piston, powder bed, and part are lowered, and the next layer of powder is spread. The drop-spread-print cycle is repeated until the entire part is completed. Removal of the unbound powder reveals the fabricated part. The local composition can be manipulated by specifying the appropriate printhead to deposit the predetermined volume of the appropriate binder.

The local microstructure can be controlled by altering the printing parameters during fabrication (24). The incorporation of microchannels effectively distributes additional seeding surfaces throughout the interior of the device, increasing the effective seeding density and uniformity. Patterned surface chemistry potentially offers spatial control over cell distribution of multiple cell types. This technology is limited by the competing needs between printhead reliability and feature resolution, as small nozzles can make finer features but are more prone to clogging. Current limitation in resolution is 100 μm for one-dimensional features (e.g., width of the thinnest printable line), and 300 μm for three-dimensional features (e.g., thickness of thinnest printable vertical walls) (25).

1.1 History

A brief history of 3D printing has been given (26). Also, the history of 3D printing in healthcare has been documented (27). The concept of three-dimensional (3D) printing goes back to Charles W. Hull (28). In 1984 Hull used UV light to harden tabletop coatings and created the term stereolithography (29). This apparatus was the world's first 3D printer. There are monographs dealing with the technology of practical 3D printers (30).

The term *3D printing* originally referred to a process that deposits a binder material onto a powder bed with inkjet printer heads layer by layer (1). More recently, the term is being used in popular vernacular to encompass a wider variety of additive manufacturing techniques. United States and global technical standards use the official term additive manufacturing for this broader sense, since the final goal of additive manufacturing is to achieve mass production, which greatly differs from 3D printing for rapid prototyping (1).

Three-dimensional printing was invented at the Massachusetts Institute of Technology (25). The history of the development of 3D printing is summarized in Table 1.1.

Additive manufacturing of polymer-fiber composites has transformed additive manufacturing into a robust manufacturing paradigm and enabled the production of highly customized parts with significantly improved mechanical properties, compared to non-reinforced polymers (33). Almost all commercially available additive manufacturing methods have benefited from various fiber reinforcement techniques.

The recent developments in 3D printing methods of fiber reinforced polymers, i.e., fused deposition modeling (FDM), laminated object manufacturing, stereolithography, extrusion, and selective laser sintering (SLS), have been reviewed (33).

Table 1.1 History of 3D printing (1,32).

Year	Event 1860–1993
1860	The photosculpture method of François Willème captures an object in 3 dimensions using cameras surrounding the subject.
1892	Blanther proposes a layering method for producing topographical maps.
1972	Mastubara of Mitsbushi Motors proposes that photo-hardened materials (photopolymers) be used to produce layered parts.
1981	Hideo Kodama of Nagoya Municipal Industrial Research Institute publishes the first account of a working photopolymer rapid prototyping system.
1981	In 1981, Hideo Kodama of Nagoya Municipal Industrial Research Institute invented two additive methods for fabricating three-dimensional plastic models (31).
1984	Charles Hull (founder of 3D systems) invents stereolithography.
1984	On 16 July 1984, Alain Le Méhauté, Olivier de Witte, and Jean Claude André filed their patent for the stereolithography process.
1984	Three weeks later in 1984, Chuck Hull of 3D Systems Corporation filed his own patent for a stereolithography fabrication system.
1988	Invention of fused deposition modeling developed by S. Scott Crump
1991	Stratasys produces the world's first FDM (fused deposition modeling) machine. This technology uses plastic and an extruder to deposit layers on a print bed.
1992	3D systems produce the first SLA 3D printing machine.
1992	DTM produces the first SLS (selective laser sintering) machine. This machine is similar to SLA technology but uses a powder (and laser) instead of a liquid.
1993	3D printing commercialized by Soligen Technologies, Extrude Hone Corporation, and Z Corporation.
1993	Dot-on-dot technique introduced by Solidscape.

Table 1.1 (cont.) History of 3D printing (1,32).

Year	Event 1995–2013
1995	Fraunhofer Institute developed the selective laser melting process.
1997	Aeromet invents laser additive manufacturing.
1999	Scientists manage to grow organs from patient's cells and use a 3D printed scaffold to support them.
2000	The first 3D inkjet printer is produced by Objet Geometries.
2000	The first multicolor 3D printer made by Z Corp.
2001	The first desktop 3D printer made by Solidimension.
2002	A 3D printed miniature kidney is manufactured. Scientists aim to produce full-sized, working organs.
2005	The RepRap project was founded by Dr. Adrian Bowyer at the University of Bath. The project was intended as a democratization of 3D printing technology. The open source hardware for the 3D printer can produce a large number of its own parts.
2008	The first 3D prosthetic leg is produced.
2008	The first biocompatible FDM material was produced by Stratasys.
2008	The RepRap Darwin is the first 3D printer to be able to produce many of its own parts.
2009	Expiration date of fused deposition modeling printing process patents.
2009	MakerBot produces a RepRap evolved kit for a wider audience.
2009	The first 3D printed blood vessel is produced by Organovo.
2011	The first 3D printed car is produced (Urbee by Kor Ecologic).
2012	The first 3D printed jaw is produced in Holland by LayerWise.
2013	Cody Wilson of Defense Distributed is asked to remove designs for the world's first 3D printed gun and the domain is seized.

In addition to extra strength, fibers have also been used in 4D printing to control and manipulate the change of shape or swelling after 3D printing, right out of the printing bed (33). Although additive manufacturing of fiber-polymer composites is increasingly being developed and is under intense scrutiny, there are some issues that need to be addressed, including void formation, poor adhesion of fibers and matrix, blockage due to filler inclusion, increased curing time, modeling, and simulation.

1.1.1 *Recently Developed Materials for 3D Printing*

According to International Data Corporation (IDC), the global amount spent on 3D printing technologies is expected to reach nearly $12 billion in 2018 (34). As 3D printers become more popular in the enterprise, there is an increased need for materials that can be used throughout the entire product development cycle.

The use of improved-performance thermoplastics in 3D printing has potential in almost all manufacturing environments. It has major uses especially in electronics, automotive and aerospace industries.

High-end 3D printing of polymers is seeing an important convergence between additive manufacturing and specialty polymers technology. This requires a collaboration between the manufacturers of 3D printing machines and speciality materials suppliers (34).

Several recently developed materials and their manufacturers have been reviewed (34,35).

These materials include poly(ether ether ketone) and PPSU filaments, nanodiamonds, and others (34).

1.1.2 *Shrinkage Compensation*

To predict the final product shape in 3D printing, finite element analysis (FEA) can be employed, for example, to simulate the structural shrinkage using a linear elastic model (36), or the complete photopolymerization, mass, and heat transfer process through a comprehensive kinetic model (37).

However, the finite element analysis method may be limited by inadequate physical understanding and a trade-off between accuracy and computational complexity (38). In addition, a large number of model parameters can be difficult to acquire accurately in practice and model complexity can reduce its practicality in direct and efficient control of shape accuracy.

Empirical models have also been developed to reduce the shrinkage through optimization of process parameters such as light intensity, exposure time, and layer thickness. Response surface modeling was adopted to optimize shrinkage at different directions (39), or to optimize the building parameters to achieve the trade-off between accuracy, building speed, and surface finish (40).

Designed experiments were used to decrease distortion and increase flatness (41). However, this approach may only control or reduce the average shape shrinkage (38).

Controlling the detailed features along the boundary of the printed product changed the CAD design to compensate for shrinkage, and polynomial regression models are used to analyze the shrinkage in X, Y, and Z directions separately (42,43). However, the prediction of deformation based on the shift of individual points can be independent of the geometry of the product, which may not be consistent with the physical manufacturing process (38).

To summarize, part shape deformation due to material shrinkage has long been studied, e.g., in casting and injection molding processes. Strategies and methods that have been developed to prescale design parts for shrinkage compensation can be classified as follows (38):

1. Machine calibration through building test parts: Similar to the calibration of computer numerically controlled machines, the additive manufacturing machine accuracy in x, y, z directions can be calibrated through building test cases (42–44).
   The dimensional accuracy of the additive manufacturing products is anticipated to be ensured during full production. However, the position of additive manufacturing light exposure may not play the same dominant role as the tool tip position of computer numerically controlled machines. The part geometry and shape, process planning, materials, and processing techniques jointly can have complex effects on the profile accuracy. The calibration of the additive manufacturing machine can therefore mostly be limited to the

scope of a family of products, specific types of materials and machines, and process planning methods.

2. Part geometry calibration through extensive trial and error build: Besides machine calibration, another strategy is to apply either a shrinkage compensation factor uniformly to the entire product or different factors to the computer-aided design for each section of a product (45).
   However, it can be time-consuming to establish a library of compensation factors for all part shapes. The library may therefore not be inclusive. In addition, interactions between different shapes or sections may not be considered in this approach. Preliminary research shows that the strategy of applying section-wise compensation may have detrimental effects on the overall shape due to *carryover effects* or interference between adjacent sections.
3. Simulation study based on first principles: Theoretical models for predicting shrinkage could potentially reduce experimental efforts. Models have been developed, e.g., in a powder sintering process (46, 47) and in metal injection molding (48). Although numerical finite element model simulation can be developed to calculate the impact of shrinkage compensation, three-dimensional deformations and distortions in additive manufacturing processes can still be rather complicated. Improving part accuracy based purely on such simulation approaches can be far from effective, and may seldom be used in practice (49).

A data processing system has been presented that may minimize errors caused by material phase change shrinkage during additive layer 3D printing (38).

Information indicative of the shape of a layer of a 3D object that is to be printed may be received. A shape that most closely corresponds to the shape of the layer may be selected from a library of shapes. Each shape in the library may have shrinkage information associated with it that includes, for each of multiple points that define a perimeter of the library shape, a radial distance to the point from an origin of a coordinate system, an angle the radial distance makes with respect to an axis of the coordinate system, and information indicative of an anticipated amount by which the point will deviate from its specified location when the shape is printed due to shrinkage.

The closeness between two shapes may be measured by the L_2 distance between the multiple points on the perimeters of two shapes. Compensation for anticipated shrinkage of the layer when printed may be calculated based on the shrinkage information that is associated with the selected shape from the library. The information indicative of the shape of the layer to be printed may be modified to minimize errors cause by shrinkage of the layer when printed based on the calculated compensation.

After the layer is printed using the selected shape with the modified shape information, error information from a user indicative of one or more size errors in the layer caused by shrinkage may be received. A new shape that is closer to the shape of the layer than the selected shape may be created and added to the library based on the selected shape and the error information from the user.

The shrinkage information in the library with the new shape may include, for each of multiple points that define a perimeter of the new shape, a radial distance to the point from an origin of a coordinate system, an angle the radial distance makes with respect to an axis of the coordinate system, and information indicative of an anticipated amount by which the point will deviate from its specified location when the shape is printed due to shrinkage.

The radial distances of at least one shape in the library may not all have a common origin. For each common origin, however, there may only be a single point at each angle.

The shrinkage information for at least one of the points in at least one of the shapes in the library may include a location-dependent and a location-independent component.

Calculating compensation for anticipated shrinkage may include computing a Taylor series expansion of the shrinkage information that is associated with the selected shape from the library. Calculating compensation for anticipated shrinkage may include calculating compensation for each point in the selected shape. A deviation may be determined between each point that defines a perimeter of the selected shape and a corresponding point on the to-be-printed layer. For each point that defines a perimeter of the selected library shape, the information of the anticipated amount by which the point will deviate from its specified location may be adjusted to include the determined deviation between the point and the corresponding point on the to-be-printed layer.

A non-transitory, tangible, computer-readable storage medium containing a program of instructions may cause a computer system running the program of instructions to implement all or any sub-combination of the functions of the data processing system that are described herein (38).

1.2 Basic Principles

The use of an inkjet-type printhead to deliver a liquid or colloidal binder material to layers of a powdered build material is involved in 3D printing (50). The printing technique involves applying a layer of a powdered build material to a surface typically using a roller. After the build material is applied to the surface, the printhead delivers the liquid binder to predetermined areas of the layer of material.

The binder infiltrates the material and reacts with the powder, causing the layer to solidify in the printed areas by, for example, activating an adhesive in the powder. The binder also penetrates into the underlying layers, producing interlayer bonding. After the first cross-sectional portion is formed, the previous steps are repeated, building successive cross-sectional portions until the final object is formed.

The apparatus for carrying out 3D printing typically move the printheads over the print surface in raster fashion along orthogonal X and Y axes. In addition to the time spent printing, each printhead move requires time for acceleration, deceleration, and returning the printhead to the starting position of the next move (50).

In design-related fields, 3D printing is used for visualization, demonstration and mechanical prototyping. It may also be useful for making patterns for molding processes. In addition, 3D printing is useful in the field of medicine (51).

The 3D printing process can be quicker and less expensive than conventional machining of prototype parts or production of cast or molded parts by conventional hard or soft tooling techniques (51).

1.2.1 4D Printing

With its additional dimension, 4D printing is emerging as a novel technique to enable configuration switching in 3D printed items (52).

Four major approaches, i.e., self-assembly of elements, deformation mismatch, bi-stability, and the shape memory effect, were identified as the generic approaches to achieve 4D printing.

The main features of these approaches were briefly discussed (52). Utilizing these approaches either individually or in a combined manner, the potential of 4D printing to reshape product design has been demonstrated by a few example applications.

1.3 Uses and Applications

1.3.1 Heat Exchangers

An overview of the most common polymer additive manufacturing processes has been presented, including vat photopolymerization, material jetting, sheet lamination, powder bed fusion, and fused filament fabrication (53).

The general strengths and challenges of the common methods were discussed (53). In particular, methods to increase the thermal performance of polymers used with the various manufacturing methods have been highlighted.

Heat exchangers enabled by polymer additive manufacturing were reviewed to assess novel designs in metal, ceramic, and polymer heat exchangers which can be made possible by the unique properties of certain polymers and the advantages offered by additive manufacturing (53).

1.3.2 3D Plastic Model

A method for the automatic fabrication of a 3D plastic model has been described (31).

The solid model is fabricated by exposing a liquid photocurable polymer of 2 *mm* thickness to UV radiation, and subsequently stacking the cross-sectional solidified layers.

1.3.3 *Gradient Refractive Index Lenses*

Gradient refractive index (GRIN) optical structures are composed of an optical material whose index of refraction, n, varies along a spatial gradient in the axial and/or radial directions of the lens (54). They have many useful applications such as making compact lenses with flat surfaces.

There are several known techniques for fabricating GRIN lenses. One approach is to press films of widely varying refractive indices together into a lens using a mold, e.g., as described before (55). This process, however, is expensive to develop (54).

A second approach for fabricating GRIN lenses is to infuse glass with ions at varying density. This approach has reached commercial production, but it is also expensive and effectively limited to small radially symmetric lenses by the depth to which ions will diffuse into glass.

A third approach for fabricating GRIN lenses is to use 3D printing technology with inks composed of a polymer matrix doped with particles which change the index of refraction of the matrix. Each printed droplet has a distinct refractive index controlled by the concentration of dopants in the polymer material. This approach has been described (56,57).

The key physical characteristics of matrix materials and dopants have been specified that are sufficient to provide all the important properties suitable for 3D printing of high quality GRIN lenses (54). Also, a variety of specific examples of such ink compounds have been detailed. These inks have the following key physical characteristics:

The matrix material is a monomer that is UV crosslinkable with 20% or less shrinkage to minimize the strain and subsequent deformation of the optical structure. The matrix material has a transmittance of at least 90% (preferably at least 99%) at the wavelengths of interest, and the viscosity of the matrix in its monomer form is less than 20 *cPoise* so that it can be inkjet printed. The matrix material is doped with nanocrystal nanoparticles at a loading of at least 2% by volume. The nanocrystals are selected such that a difference in index of refraction between the doped and undoped matrix material is at least 0.02, i.e., $\Delta n \geq 0.02$.

The nanocrystal sizes are sufficiently small that they do not induce

Mie or Rayleigh scattering at the wavelengths of interest (e.g., less than 50 *nm* in size for visible wavelengths, less than 100 *nm* for IR wavelengths). The nanocrystal material, as well as the doped matrix material, preferably has a transmittance of at least 90% (more preferably, at least 99%) in a predetermined optical wavelength range (e.g., visible spectrum). The nanocrystals are functionalized with silane ligands. In some embodiments, the ligands are less than 1.2 *nm* as measured radially from the nanoparticle core. Some of the ligands attach to the nanoparticle core at their anchor end and have a buoy end. The buoy end is either reactive to the monomer or non-reactive to the monomer. A plurality of ligands can be used with different functionalization based on anchor and buoy ends.

The matrix or host polymer is 1,6-hexanediol diacrylate, and the nanoparticle is an organometallic compound. 1,6-Hexanediol diacrylate is a well-known material for the fabrication of clear coatings. It has a low viscosity of 7.9 *cP* making it amenable towards dispensing using drop-on-demand techniques such as inkjet printing. It also has a large spectral window in which greater than 99% transparency is observed, making it ideal to construct lensing material. 1,6-Hexanediol diacrylate also has an index of refraction of 1.456. By using an organometallic compound having an index of refraction different from that of 1,6-hexanediol diacrylate, using drop-on-demand techniques such as inkjet printing, gradient refractive index lenses may be fabricated having control of the index of refraction in three dimensions.

Ligand functionalization of clear, transparent metallic salts provide matrix compatibility with 1,6-hexanediol diacrylate, allowing high density loading of the organometallic salt into the matrix. Furthermore, due to a difference of index of refraction between undoped 1,6-hexanediol diacrylate and 1,6-hexanediol diacrylate doped with the functionalized metallic salt, GRIN lenses may be formed using drop-on-demand printing techniques such as inkjet printing. The metallic salts interact favorably with a host matrix material such that greater than 90% transparency is obtained in the spectral region spanning 375 *nm* through 1600 *nm* (54).

1.3.4 *Photoformable Composition*

A method for fabricating an integral three-dimensional object from layers of a photoformable composition has been presented (58).

A semi-permeable film can be used, which is impermeable to the photoformable composition but is permeable to a deformable-coating-mixture that is non-wetting and immiscible with the photoformable composition.

The deformable-coating-mixture passes through the membrane by diffusion and forms a thin, slippery surface on the photoformable composition side of the membrane, thereby eliminating any adhesion forces caused by chemical, mechanical or hydrogen bonds.

The following steps comprise the method for fabricating an integral three-dimensional object from successive layers of a photoformable composition (58):

1. Positioning a transparent, semi-permeable film,
2. Contacting an interface with a gaseous oxygen-containing atmosphere,
3. Allowing gaseous atmosphere to permeate through the film and also partially permeate into a photoformable composition layer,
4. Exposing the photoformable layer to radiation imagewise through the film, thus making a photoformed layer and a release coating,
5. Sliding the formed film from the photoformed layer,
6. Positioning the film in order to form a photoformable layer between the previously made photoformed layer and the other surface, and
7. Repeating the previous steps until all the layers of the three-dimensional object are formed.

1.3.5 *Comb Polymers*

Suspensions stabilized by comb polymers may serve as colloid-based inks for fabricating three-dimensional structures.

In such applications, the improved dispersion of nanoparticles that is yielded by comb polymers has proven to be advantageous (59,60).

Such inks enable the production of three-dimensional structures with feature sizes as small as 100 μm (61).

Preferred comb polymers contain two types of side chains and are water soluble. The first type of the side chain has moieties that ionize at the pH of the colloidal suspension. The second type of side chain is nonionizable. Comb polymers from poly(acrylic acid) and poly(ethylene oxide) have nonionizable side chains, in addition to ionizable side chains (62).

1.3.6 *Post-Processing Infiltration*

Printing processes may include a post-processing infiltration step in order to increase the strength of the printed article using two-component casting resins or adhesives or one-component cyanoacrylate adhesives to achieve greater durability in a three-dimensional article (63).

Furthermore, they may be infiltrated with a liquid plasticizer to obtain strengths comparable to that of articles formed with cyanoacrylate adhesive.

The usage of infiltrant materials for plasticized sintering may provide some advantages over conventional methods. The plasticizer may be ethanol, benzene sulfonamide or propylene carbonate. These compounds are shown in Figure 1.1.

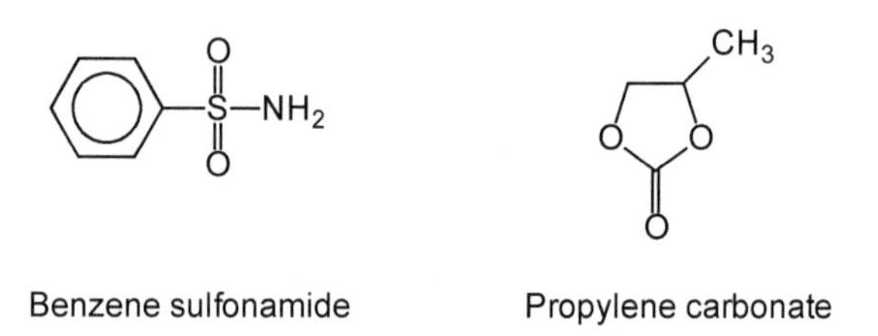

Figure 1.1 Plasticizers.

An extremely high solubility of the polymer in plasticizer may be undesirable because it may result in over-plasticization or the dissolution of the thermoplastic additive. In that case, the glass transition temperature may be close or below room temperature, which may cause distortion and weak particle bonding. The plasticizer material may be preferably selected from materials that have low solubility at room temperature but greater solubility at higher temperatures. To reduce the solubility, the plasticizer may be diluted either by a solvent that is removed after sintering or an inert solid material that may remain in the three-dimensional article after cooling.

Using two-component casting resins, such as epoxy-amine, isocyanate-amines, or isocyanate-polyol systems, decreases the ease of use for the end-user by incorporating extra mixing steps, imposing pot-life constraints, and giving rise to safety, health, and environmental issues.

One-component cyanoacrylate adhesives typically offer better ease of use because these materials do not require mixing, but they may create safety, health, and environmental issues such as fumes, irritation, and adhesion to skin, and may not be stable when exposed to the open atmosphere for long periods of time.

The plasticized assisted sintering of a material consisting of a thermoplastic particulate increases ease of use by offering a method in which the process can be automated, whereby the article is immersed in a stable, one-component liquid medium for a predetermined amount of time and allowed to cool to a handling temperature (63).

The steps of a plasticizer-assisted sintering process are (63):

1. A layer is formed of a substantially dry particulate material containing thermoplastic particulate, a plaster, or a water-soluble polymer such as a water-soluble adhesive,
2. An aqueous fluid binder is applied to the layer of the dry particulate material in a predetermined pattern to cause binding in the areas to which the binder is applied,
3. The previous steps are repeated sequentially to form a three-dimensional article,
4. After complete setting of the polymer, the three-dimensional article is removed from the build.
5. Then the three-dimensional article is submerged in the plasticizer solution, and
6. Optionally, the particulate material may be exposed to additional energy in the form of conventional heat, visible or infrared light, microwave, or radio frequency, for additional sintering of particulate material.

1.3.7 *Sensors and Biosensors*

Applications of 3D printing in analytical and bioanalytical chemistry have been on the rise, with microfluidics being one of the most represented areas of 3D printing in this chemistry branch (64).

Most stages of the analytical workflow comprising sample collection, pretreatment and readout, have been enabled by 3D printed components. Sensor fabrication for detecting explosives and nerve agents, the construction of microfluidic platforms for pharmacokinetic profiling, bacterial separation and genotoxicity screening, the assembly of parts for on-site equipment for nucleic acid-based detection and the manufacturing of an online device for *in-vivo* detection of metabolites are just a few examples of how additive manufacturing technologies have aided the field of (bio)analytical chemistry.

The most relevant trends of 3D printing applications in the abovementioned fields have been reviewed (64).

1.3.7.1 Fluidic Control

The synthesis and assembly of a diverse range of materials, including spray-based synthesis of inorganic nanoparticles and conductive filaments, extrusion-based fabrication of hydrogel fibers and sheets, and the preparation of composite solid films have been reported (65). In the reported studies the properties of these composites were examined and potential applications of these materials were shown. Also, the advantages of material fabrication in 3D printed microfluidic devices were highlighted.

The most representative use of 3D printing technologies in bioanalytical chemistry is based on fluidic control. In 2015, Gowers *et al.* developed a flow microfluidic device for *in-vivo* monitoring of lactate and glucose in people while cycling (66). They printed a microfluidic chip with channels in the range of 520 μm to 1000 μm in height and 520 μm to 550 μm in width, using 3SP technology (photopolymerization). The metabolites were extracted directly from the individuals with a FDA-approved dialysis probe. The dialysate liquid was pumped into the microfluidic chip where two needle electrodes, each functionalized with the specific enzyme for each analyte, were held within a soft structure printed with a multimaterial stereolithography (SLA) printer (64). This method allowed the fabrication of rigid and soft materials necessary to seal the electrodes inside the chip, preventing any sample leakage, enabling real-time measurement of metabolites related to exercise performance.

Another functional design was proposed by Lee *et al.* (67), who reported a 3D printed helical microchannel to detect pathogenic bacteria in milk. *Escherichia coli* cells were magnetically trapped using antibody-coated magnetic nanoparticle clusters, followed by the separation of the free particles and bacteria-loaded particles using the helical 3D printed microfluidic device. By creating a trapezoidal geometry inside the microfluidic channels (1000 μm in width, 250 μm and 500 μm in inner and outer heights, respectively) a 3D-printed microfluidic device was formed. With the help of a sheath flow (inner outlet), the bacteria-loaded particles accumulated near the inner fluid wall created by the sheath flow while the unloaded particles, being smaller, stayed in the outer wall.

This way, the *E. coli*-containing particles were released at the inner outlet for subsequent spectrophotometric detection, while the bacteria-free particles leaked through the outer outlet. Thanks to this carefully designed structure, the measured absorbance was highly specific for the targeted microorganism, without the need of using labels or intermediate signal-generating species. The device was printed using a SLA printer and a commercial transparent/water-resistant resin (64).

1.3.7.2 *Strain Sensors*

Conductive silicon rubbers are potential candidates for strain sensors owing to their specific electrical response and superior mechanical flexibility (68).

Carbon fiber-filled conductive silicon rubbers were printed using an extrusion device. A thixotropic agent was added to modify the mobility and the viscosity of the liquid conductive silicon rubber.

As matrix rubber, a methyl vinyl rubber was used. Castor oil served as a thixotropic agent. The preparation of the composition runs as follows (68):

Preparation 1–1: The matrix was mixed evenly with 10 *phr* (parts per hundred parts of polymer matrix), 60 *phr* carbon fiber and a certain amount of thixotropic agent sequentially. A liquid conductive rubber was obtained in this step. Then followed vacuumizing and filling. After vacuumizing the above liquid conductive rubber to ensure a void-free printed object, the printing cylinder was filled with the liquid conductive rubber. The printing process was based on an extrusion device that cooperated with a desktop printer.

It was found that a conductive silicon rubber with 5% thixotropic agent addition exhibited a good shape-retention. Fibers in matrix were observed to be oriented in the printing direction, resulting in an anisotropic electrical and mechanical behavior.

The printed conductive silicon rubbers showed better electrical and mechanical properties along the orientation direction of fibers. In particular, the volume resistivity at the orientation direction was 6.8 times lower than that at perpendicular direction. Higher tensile strength, larger elongation at break, and higher Young's modulus were found along the orientation direction when the printed conductive silicon rubbers were stretched, where a large number of fibers were pulled out and visible holes remained at the fractured surface.

The electrical responses of the conductive silicon rubbers under various loadings, including stretching, compressing, bending, twisting and cyclic folding, were closely related with deformations of the conductive silicon rubbers.

In addition, sandwich strain sensors were finally fabricated to verify a practical application as a motion sensor of the printed conductive silicon rubbers (68).

1.3.7.3 Genotoxicity

An automated array has been fabricated to assess the genotoxic potential of cigarettes, e-cigarettes and environmental samples (69). Potentially genotoxic reactions from e-cigarette vapor similar to smoke from conventional cigarettes have been demonstrated. Furthermore, untreated wastewater showed a high genotoxic potential compared to negligible values for treated wastewater from a pollution control treatment plant.

In detail, microfluidic arrays were printed from a clear acrylate resin using a SLA 3D printer. The arrays consisted of three sample chambers (17 $mm \times 5$ $mm \times 2.5$ mm in length, width and height, respectively) feeding three detection channels (23 $mm \times 3$ $mm \times$

0.65 *mm*) hosting a microwell-patterned pyrolytic graphite detection array (64).

Films consisting of a ruthenium-based metallopolymer, human liver microsome as cytochrome P450 source and calf thymus deoxyribonucleic acid (DNA) were grown via a layer-by-layer electrostatic assembly in these microwells (70).

Each array was 3D printed in less than 1 *h* at a low fabrication cost. The detection channels were designed to hold a stainless-steel wire counter electrode and a Ag/AgCl wire used as reference to make up an electrochemiluminescent detection cell controlled by means of a hand-held potentiostat.

Also, the assay exploited the natural catalytic activity of cytochrome P450 by applying a fixed potential under a sample flow (64). This property generated metabolites from the molecules present in the samples. The principle of the assay was based on the reaction of these metabolites with DNA if they had genotoxic activity, disrupting the double helix by the formation of adducts or causing strand breakage. Free and available guanine bases resulting from these genotoxic disruptions reacted with a ruthenium complex that, in an excited state, generated an electrochemiluminescent signal. This signal was proportional to the DNA damage caused by the samples (64).

1.4 Magnetic Separation

A magnetic separation principle also including a trapezoidal chamber was later utilized to detect *E. coli* through an adenosine triphosphate (ATP) luminescence assay (71) or DNA amplification (72). The 3D printed microfluidic pre-concentrating device also contains a microfiltering comb-like structure, aimed at filtering out large particles as an extra sample-cleaning step.

In the most recent report of Park *et al.*, the structures were directly 3D printed with digital light processing technology using a photocurable resin acrylate-based photopolymer without post-printing processing (71).

Another 3D printing-enabled ATP luminescence assay has been reported (73). a fluidic chip was fabricated using a SLA 3D printer and a proprietary resin-made acrylate-based and epoxy-based mixtures. The object chip contained two inlet ports with an internal radius of 500 μm and it took less than 30 min to create a very cheap prototype.

1.5 Rapid Prototyping

The field of rapid prototyping involves the production of prototype articles and small quantities of functional parts, as well as structural ceramics and ceramic shell molds for metal casting directly from computer-generated design data (51,74).

Two well-known methods for rapid prototyping include (74):

1. A selective laser sintering process and
2. A liquid binder three-dimensional printing process.

These techniques are similar as both use layering techniques to build three-dimensional articles.

Both methods form successive thin cross sections of the desired article. The individual cross sections are formed by bonding together adjacent grains of a granular material on a generally planar surface of a bed of the granular material.

Each layer is bonded to a previously formed layer to form the desired three-dimensional article at the same time as the grains of each layer are bonded together (74).

Rapid prototyping processes are layered manufacturing techniques wherein an article, e.g., metal casting mold or a prototype part, is progressively made in a series of sequentially built-up layers (75).

A classification of rapid prototyping processes is schematically shown in Figure 1.2.

The direct ink writing of three-dimensional ceramic structures has been reviewed (77). Both droplet- and filament-based direct ink writing techniques have been detailed. Various ink designs and their corresponding rheological behavior, ink deposition mechanics, potential shapes and the toolpaths required, and representative examples of 3D ceramic structures have been presented.

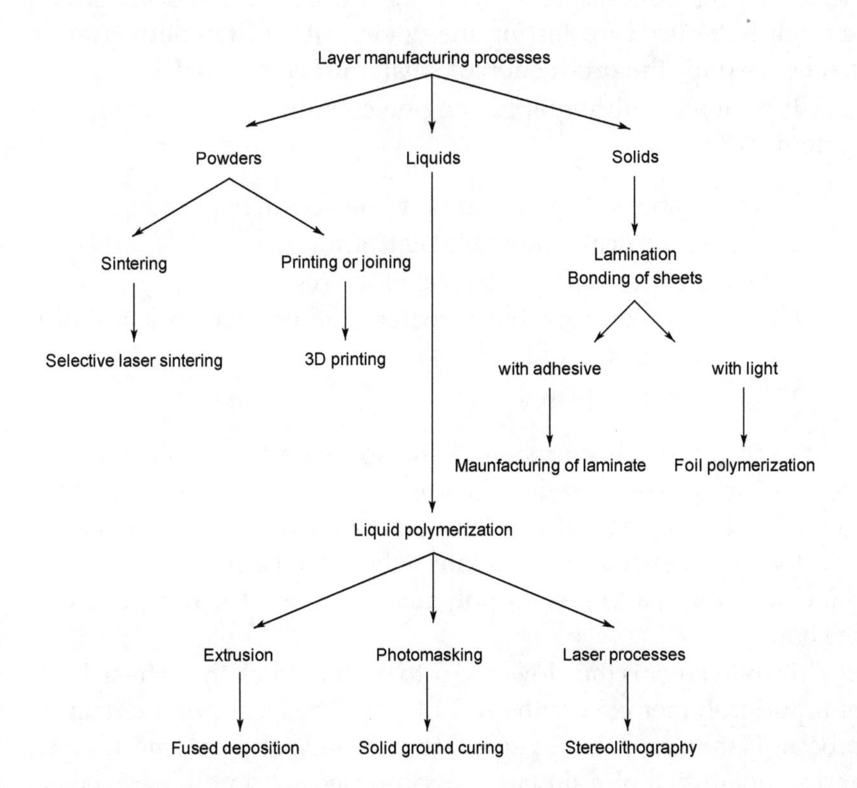

Figure 1.2 Classification of rapid prototyping processes (76).

1.5.1 Variants of Rapid Prototyping

1.5.1.1 Stereolithography

The current state-of-the-art of stereolithography has been described in detail (78). The main advantage of stereolithography in comparison to other rapid prototyping methods is the ability to fabricate final models in a very short time. In addition, the prototypes fabricated by stereolithography have tight tolerances and are strong enough to be used for testing the device. Also, stereolithography can be used for the production of small numbers of articles.

A typical stereolithographic engine contains the following subsystems (78):

1. Removable vat with a liquid photosensitive polymer,
2. Precise vertically movable platform,
3. Devices for sensing the level of the resin,
4. Horizontally movable recoater blade to create a uniform thick layer of liquid,
5. UV lamp for curing.

A variant of such a process is the so-called SLA 1 system (79). Here, a computer-controlled, focused UV laser is scanned over the top surface of a bath of a photopolymerizable liquid polymer to selectively polymerize the polymer where the laser beam strikes it, thereby forming a first solid polymeric layer on the top surface of the bath.

This solid layer is then lowered into the bath such that a fresh layer of liquid polymer covers the solid layer. The laser polymerization process is then repeated to generate a second solid polymeric layer, and so on, until a plurality of superimposed solid polymeric layers complete the desired article.

An apparatus for the production of three-dimensional objects by stereolithography has been described (29).

Three-dimensional objects are generated within a fluid medium, which is selectively cured by radiation brought to a selective focus at prescribed intersection points within the three-dimensional volume of the fluid medium.

Many liquid state chemicals are known which can be induced to change to solid-state polymer plastic by irradiation with UV light.

Such UV curable chemicals are used as ink for high speed printing and can also be used for stereolithography (29).

1.5.1.2 *Selective Laser Sintering*

The SLS process is also known as laser powder deposition (80). Laser deposition is a solid free-form fabrication method in which a laser beam is used to melt an addition material to create a material track with approximately hemispherical cross section.

By partially overlapping the consecutive tracks, continuous layers of material can be obtained, which can be used as coatings if deposited by laser cladding, or also overlapped to create 3D objects directly from their computer representation.

The development of rapid manufacturing techniques by laser deposition started in the mid-1990s. It started with the blown powder laser cladding process, a technology that was pioneered by Weerasinghe and Steen as a coating process (81).

In the SLS process, a computer-controlled laser beam sinters selected areas of multiple layers of loosely compacted powder, e.g., plastic, metal, ceramic, or wax, layer-by-layer, until the article is completely built-up. The SLS procedure has been described in more detail (79).

Still another variant is known as the 3D printing rapid prototyping process wherein a computer-controlled inkjet printing device propels a stream of binder from one or more jets onto select areas of a first layer of loose particles of some 60–140 μm in diameter according to a pattern controlled by a computer. The 3D printing rapid prototyping process has been described in more detail (82–84).

Particle layers may be formed by depositing either dry particles, or particles suspended in a volatile liquid, onto a working surface before the binder is applied. The volatile liquid is allowed to evaporate from a first layer before depositing a second layer. This process is then multiply repeated until the article is completed.

The binder in one layer is at least partially cured before the next layer of particles is laid down. The finished article may thereafter be heated for further drying or curing of the binder to provide a final article with sufficient green strength for handling.

Thereafter, the article may further be heated to sinter or to weld the particles together to form a finished, albeit porous, article.

1.5.2 3D Microfluidic Channel Systems

Topologically complex 3D microfluidic channel systems can be made in poly(dimethyl siloxane) systems. The procedure is called the membrane sandwich method (85).

A thin membrane having channel structures is molded on each face and sandwiched between two thicker, flat slabs that provide a structural support. Two masters are fabricated by rapid prototyping using two-level photolithography and replica molding and aligned face to face. The prepolymer is between these masters. Then the poly(dimethyl siloxane) is thermally cured.

This method can fabricate a membrane containing a channel that crosses over and under itself, but does not intersect itself and, therefore, can be fabricated in the form of any knot.

1.5.3 Aluminum and Magnesium Cores

It would be desirable to manufacture lightweight articles from particles made from aluminum or magnesium or their alloys using the 3D printing rapid prototyping technique (75).

However, in the past it has not been possible to do so owing to the reactivity of Al and Mg particles and their propensity to readily oxidize in air to form an oxide skin on the particle's surface that impedes sintering/welding of the particles to each other.

Coated particles were developed that are composed of a core metal selected from Al, Mg and their alloys, respectively, and a coating that protects the core from oxidation.

The coating is a metal whose oxide is reducible by heating in a non-oxidizing atmosphere: for example, the coating metal is selected from copper, nickel, zinc, or tin. Copper is the most preferred coating metal.

Either only one layer or several layers of copper are used. For example, a copper topcoat is underlaid with a first undercoating such as Zn or Si that can form an alloy with the copper and the core metal. The coating alloy melts below the liquidus temperature of the Al or Mg core metal. Furthermore, for aluminum particles, the undercoat preferably comprises Zn, Si, or Mg.

Suitable binders are polymeric resins, e.g., a butyral resin, or inorganic materials, e.g., silicates. The binders should be soluble either in water or volatile organic solvents (75).

Also, a carbohydrate-based binder has been described that contains sugars and starches (86).

1.5.4 *Cellular Composites*

Lightweight cellular composites, composed of an interconnected network of solid struts that form the edges or faces of cells, are an emerging class of high-performance structural materials. These devices may find potential application in (87):

- High stiffness sandwich panels,
- Energy absorbers,
- Catalyst supports,
- Vibration damping, and
- Insulation.

An epoxy-based ink that enables 3D printing of cellular composites has been reported (87).

This ink allows the controlled alignment of multiscale, high aspect ratio fiber reinforcement to create hierarchical structures similar to balsa wood. The 3D printing technology offers a flexibility in achieving controlled composition, geometric shape, function, and complexity.

The inks are prepared by mixing an epoxy resin (Epon 826 epoxy resin) with dimethyl methyl phosphonate, nanoclay platelets (Cloisite 30B), silicon carbide whiskers (SI-TUFF SC-050), and milled carbon fibers (Dialead K223HM). Also, an imidazole-based ionic liquid is employed as a latent curing agent (Basionics VS 03) (87).

1.5.5 *Powder Compositions*

Powder compositions have been disclosed that can be used with the traditional 3D printing technology (88). 3D printed building materials with comparable compressive strength to standard concrete but greater tensile strength have been developed.

The composition has up to four components (88):

1. An adhesive material,
2. An optional absorbent or hydrator material,
3. A base material, and
4. An optional additive that improves the strength of the final object.

The components are detailed in Table 1.2.

Table 1.2 Components for powder compositions (88).

Adhesive component	Hydrator component
Powdered sugar	Maltodextrin
Poly(vinyl alcohol)	Sodium silicate
Polyureic formaldehyde	Alkaline polyphosphates
Wheat paste	Carbonates
Methyl cellulose	Polycarbonates
Powdered wood glue	Gum arabic
Water putty	Alkaline lignosulfonates
Base component	**Reinforcing component**
Expansion cement	Reinforcement fibers
Fly ash	Sake
Patching cement	Fiber mesh
Rapid set cements	Fly ash
Portland cement	Poly(vinyl alcohol) fiber
Hydraulic cement	Carbon fiber

1.5.6 *Organopolysiloxane Compositions*

A typical method of optically manufacturing a three-dimensional object involves selectively irradiating an ultraviolet laser beam to the surface of liquid photocurable resin contained in a vat under the control of a computer to harden the photocurable resin so that a photo-cured resin layer having a predetermined thickness is obtained. Then, a layer of liquid photocurable resin is supplied onto the cured resin layer and then likewise irradiating an ultraviolet laser beam to the liquid photocurable resin layer to form a cured resin layer contiguous to the previous one, and repeating the laminating operations until a desired three-dimensional object is obtained (89).

There is a need for resins which when processed by the rapid prototyping technique, exhibit a rubber-like property. That is, having a nature which easily undergoes deformation, without rupture, under an applied stress and resumes the original shape after the stress is relieved.

A rapid prototyping resin composition of the actinic energy radiation cure type has been developed, which has improved storage stability and aging stability prior to exposure to actinic energy radiation. It experiences little viscosity buildup during long-term storage at elevated temperature and shows a high cure sensitivity to actinic energy radiation, typically light. When exposed to actinic energy radiation, it produces a cured part in a smooth and efficient manner, which has improved dimensional precision, shaping precision, water resistance, and moisture resistance. Furthermore, it exhibits a stable rubber elasticity over a long time.

It has been found that a silicone resin composition of the actinic energy radiation cure type, especially a silicone rubber, i.e., a organopolysiloxane elastomer, based material containing an alkenyl-containing organopolysiloxane, a mercapto-containing organopolysiloxane, and an alkenyl-containing resin with a three-dimensional network structure is an effective rapid prototyping resin.

This resin composition can be obtained by blending an actinic energy radiation-sensitive polymerization initiator and an actinic energy radiation absorber. Although the resin composition has a high cure sensitivity to actinic energy radiation and rapidly cures when exposed to actinic energy radiation, the resin composition is easy to handle in that it has an improved storage stability and aging stability. When it is stored for a long time, even at elevated temperature, it experiences only a little viscosity buildup and maintains a flowable state compatible with rapid prototyping.

Typical radical polymerization initiators include benzyl and dialkylacetal derivatives thereof, acetophenone compounds, benzoin and alkyl ether derivatives thereof, benzophenone compounds, and thioxanthone compounds. 2-Hydroxy-2-methyl-1-phenyl-propan-1-one is most preferred, because it is liquid at normal temperature, readily dissolvable or dispersible in the silicone composition.

Actinic energy radiation absorbers that can be used are benzotriazole compounds, benzophenone compounds, phenyl salicylate compounds and cyanoacrylate compounds. Examples are collected in Table 1.3. Some compounds are shown in Figure 1.3.

Table 1.3 Actinic energy radiation absorbers (89).

Compound
2-(2′-Hydroxy-5′-methylphenyl)benzotriazole
2-[2′-Hydroxy-3′-butyl-5′-(2″-carboxyoctylethyl)phenyl]-benzotriazole
2-Hydroxy-4-methoxybenzophenone
p-Methylphenyl salicylate
2-Hydroxy-4-octyloxybenzophenone
2-(2′-Hydroxy-3′-*tert*-butyl-5′-methylphenyl)-5-chlorobenzotriazole
2-(2′-Hydroxy-3′,5′-di-*tert*-butylphenyl)-5-chlorobenzotriazole
2-(2′-Hydroxy-5′-*tert*-octylphenyl)benzotriazole
Benzenepropanoic acid

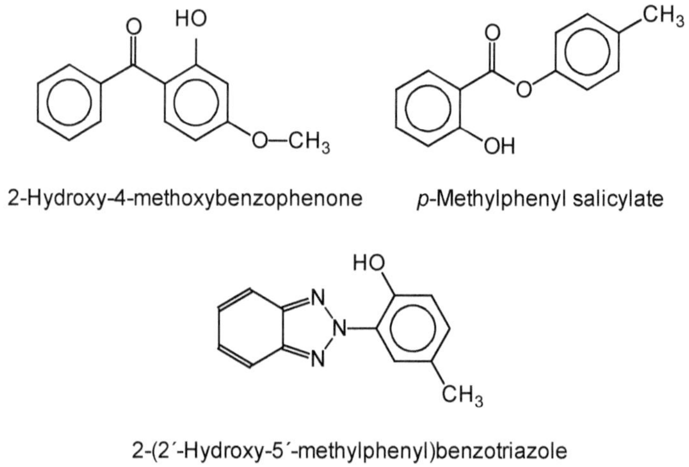

2-Hydroxy-4-methoxybenzophenone

p-Methylphenyl salicylate

2-(2′-Hydroxy-5′-methylphenyl)benzotriazole

Figure 1.3 Actinic energy radiation absorbers.

Several examples for such compositions have been reported in detail (89).

1.5.7 *Thermoplastic Powder Material*

Thermoplastic materials are widely used for engineering and consumer products. These materials, therefore, are particularly attractive for prototyping, because they are typically also used in the final manufacturing method.

A thermoplastic powder has been adapted for 3D printing. The powder includes a blend of a thermoplastic particulate material, and an adhesive particulate material, with the adhesive particulate material being adapted to bond the thermoplastic particulate material when a fluid activates the adhesive particulate material. Thermoplastic materials are collected in Table 1.4. Adhesive particulate materials are collected in Table 1.5. Adhesives are collected in Table 1.6.

1.5.8 *Plasticizer-Assisted Sintering*

A strong printed article may be made by 3D printing over a substantially dry particulate material including an aqueous-insoluble thermoplastic particulate material (63). The printed article is further post-processed by infiltrating a liquid medium into the article. The liquid medium selectively plasticizes the aqueous-insoluble thermoplastic particulate material, lowering the glass transition temperature of the material. This facilitates the sintering of the thermoplastic particulate material to bond together the matrix of the article, thereby increasing the article's durability.

An infiltrant comprises a hydroxylated hydrocarbon, a wax, a plasticizer, and a stabilizer, with the aqueous-insoluble thermoplastic particulate material being plasticized by the infiltrant.

1.5.9 *Radiation-Curable Resin Composition*

In order to reduce the time to build a part via a stereolithography process, modern stereolithography machines require a more versatile liquid radiation-curable resin composition.

A liquid radiation-curable resin capable of curing into a solid upon irradiation has been described that contains a cycloaliphatic epoxide with a linking ester group, oxetanes, polyols, methacrylate components, impact modifiers and photoinitiators (90).

Table 1.4 Thermoplastic powder materials (74).

Polymer type
Acetal poly(oxymethylene)
Poly(lactide)
Poly(ethylene)
Poly(propylene)
Ethylene vinyl acetate
Poly(phenylene ether)
Poly(vinylidene fluoride)
Poly(etherketone)
Poly(butylene terephthalate)
Poly(ethylene terephthalate)
Poly(cyclohexylenemethylene terephthalate)
Poly(phenylene sulfide)
Poly(methylmethacrylate)
Poly(sulfone)
Poly(ethersulfone)
Poly(phenylsulfone)
Poly(acrylonitrile)
Poly(acrylonitrile-butadiene-styrene)
Poly(amide)
Poly(condensates of urea-formaldehyde)
Poly(styrene)
Poly(olefin)
Poly(vinyl butyral)
Poly(carbonate)
Poly(vinyl chlorides)
Poly(ethylene terephthalate)
Ethyl cellulose
Hydroxyethyl cellulose
Hydroxypropyl cellulose
Methyl cellulose
Cellulose acetate
Hydroxypropylmethyl cellulose
Hydroxybutylmethyl cellulose
Hydroxyethylmethyl cellulose
Ethylhydroxyethyl cellulose
Cellulose xanthate

Table 1.5 Adhesive particulate materials (74).

Adhesive type
Poly(vinyl alcohol)
Sulfonated polyester polymer
Sulfonated poly(styrene)
Octylacrylamide/acrylate/butylaminoethyl methacrylate copolymer
Acrylates/octylacrylamide copolymer
Poly(acrylic acid)
Poly(vinyl pyrrolidone)
Styrenated poly(acrylic acid)
Poly(ethylene oxide)
Sodium poly(acrylate)
Sodium poly(acrylate) copolymer with maleic acid
Poly(vinyl pyrrolidone) copolymer with vinyl acetate
Butylated poly(vinyl pyrrolidone)
Poly(vinyl alcohol)-*co*-vinyl acetate
Starch
Modified starch
Cationic starch
Pregelatinized starch
Pregelatinized modified starch
Pregelatinized cationic starch

Table 1.6 Adhesives (74).

Adhesive type
Plaster
Bentonite
Precipitated sodium silicate
Amorphous precipitated silica
Amorphous precipitated calcium silicate
Amorphous precipitated magnesium silicate
Amorphous precipitated lithium silicate
Portland cement
Magnesium phosphate cement
Magnesium oxychloride cement
Magnesium oxysulfate cement
Zinc phosphate cement
Zinc oxide-eugenol cement
Aluminum hydroxide
Magnesium hydroxide
Calcium phosphate
Sand
Wollastonite
Dolomite

1.6 Solution Mask Liquid Lithography

A novel methodology for printing 3D objects with spatially resolved mechanical and chemical properties has been described (91).

Photochromic molecules are used to control the polymerization reaction through coherent bleaching fronts, providing large depths of cure and rapid build rates without the need for moving parts. The coupling of these photoswitches with resin mixtures containing orthogonal photocrosslinking systems allows the simultaneous and selective curing of multiple networks, providing access to 3D objects with chemically and mechanically distinct domains.

It was observed that irradiating optically dense photochromic solutions with collimated light leads to rapid and linear photobleaching fronts (92–94)

In order to demonstrate the performance of solution mask liquid lithography, a resin with *N,N*-dimethyl acrylamide as the monomer, 1,4-butanediol diacrylate as the crosslinking agent, a xanthene derivative as the photosensitizer, and a diarylethene photoswitch, i.e, 1,2-bis(2-methyl-1-benzothiophen-3-yl)perfluorocyclopentene, cf. Figure 1.4, as the mask was prepared and exposed to collimated green light at around 530 *nm*.

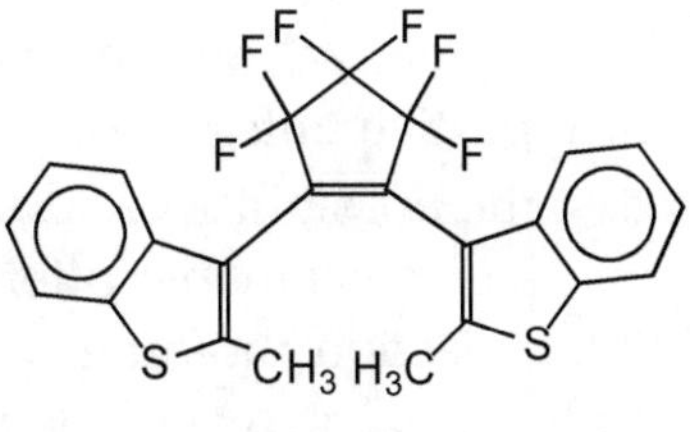

F F F F F F S CH_3 H_3C S

Figure 1.4 1,2-Bis(2-methyl-1-benzothiophen-3-yl)perfluorocyclopentene.

The front propagates through the resin with time in a linear fashion. This provides a large depth of cure, greater than 6 *cm*. This is in direct contrast to the uncontrolled polymerization observed in the absence of any dye or thin cures due to traditional, nonbleaching dyes when exposed to 530 *nm* light.

Spatial confinement in the lateral dimension with solution mask liquid lithography was further illustrated by exposure of a resin to collimated green light through a focusing lens. Significantly, curing traced the focal envelope of the lens, leading to the formation of a cone-shaped 3D object. The growth of this cone, 2.2 *cm* in height, requires no moving parts and occurs with a build rate of 50 $cm\,h^{-1}$, using a narrow-band collimated green LED with 450 $mW\,cm^{-2}$.

To print a cone of the same dimensions via a state-of-the-art UV-based inkjet printing methodology requires a build time of greater than 1 *h* at a build rate of 2 $cm\,h^{-1}$.

In addition, pairing solution mask liquid lithography resins with a projector enabled the production of parts with a lateral resolution of approximately 100 μm, on par with commercial printers.

This comparison clearly demonstrated the versatility of solution mask liquid lithography as a rapid, low-cost continuous 3D printing process (91).

1.7 Vat Polymerization

Vat polymerization (95,96) is also known as digital light processing. Photopolymerization processes make use of liquid, radiation-curable resins, or photopolymers, as their primary materials. Most photopolymers react to radiation in the UV range of wavelengths, but some visible light systems are used as well. Upon irradiation, these materials undergo a chemical reaction to become solid. This reaction is called photopolymerization, and is typically complex, involving many chemical participants (96).

It has been shown that the possibility of additive manufacturing is a cost-efficient, quick and sustainable alternative for injection molding inserts (97,98). Injection molding requires the cavity to withstand pressure and heat cycles during the injection process.

Advantages are the smaller voxel size which is achievable by vat polymerization compared to subtractive manufacturing technologies. This applies especially for corners and sharp edges as well as micro-features (99).

A HTM 140 photopolymer (100) was chosen with respect to the thermal and mechanical properties to support the fiber structure as a matrix material and to allow a comparison for the vat polymerization machining.

The insert production was performed by a vat polymerization machine, exposing a resin vat filled with the above-mentioned photopolymer from the bottom. The light intensity was increased by 10% above the normal exposure value in order to overthrow the obscuring of the resin by the carbon fibers. The layer thickness of 35 μm was set to support the increased density of the photopolymer and carbon fibers. In order to provide an even fiber distribution, the build plate was moved up by 5 mm before increasing the layer height for the next exposure level. This step was performed to allow a fiber flow below the build plate and to avoid a clustering of the fibers at the corners of the build plate. A flexible self-peeling vat was used during the process, reducing the forces induced in the part during the lifting process.

Additively manufactured injection molding inserts were shown to be an environmentally friendly, cheap and fast method for flexible rapid prototyping and pilot production in injection molding technology despite the low lifetime. Fiber reinforced polymers in the additive manufacturing technology helped to improve the lifetime significantly and reduce crack propagation velocity to a minimum.

Short carbon fibers showed an increasing effect on Young's modulus, but had a decreasing effect on break strength and tensile strength, allowing the efficient use of the material in low-strength regimes where low deformation is needed during a stress-inducing process. Despite the average increase of Young's modulus, it was shown that the fiber-matrix interface requires improvement in order to reduce failures in the part, therefore also reducing standard deviation of the mechanical properties such as Young's modulus, break strength and tensile strength (99).

1.7.1 Poly(dimethyl siloxane)-Based Photopolymer

Vat photopolymerization, also referred to as stereolithography, is widely considered the most accurate and highest-resolution additive manufacturing technique (96). Vat polymerization uses a vat of liquid photopolymer resin, out of which the model is constructed layer by layer (101). An ultraviolet (UV) light is used to cure or harden the resin where required, whilst a platform moves the object being made downwards after each new layer is cured.

A poly(dimethyl siloxane)-based photopolymer was described that exhibits simultaneous linear chain extension and crosslinking (102). This material was suitable for vat photopolymerization additive manufacturing.

The photopolymer compositions consisted of dithiol and diacrylate functional poly(dimethyl siloxane) oligomers, where simultaneous thiol-ene coupling and free-radical polymerization provided for linear chain extension and crosslinking, respectively (102).

A typical synthesis of an acrylamide-terminated poly(dimethyl siloxane) was done according to standard Schotten-Baumann conditions (103). The Schotten-Baumann reaction is a method to synthesize amides from amines and acid chlorides (104).

Furthermore, a thiol-terminated poly(dimethyl siloxane) was synthesized using standard Fischer esterification conditions. Fischer esterification is a special type of esterification by refluxing a carboxylic acid and an alcohol in the presence of an acid catalyst (105).

For photocuring, diphenyl(2,4,6-trimethylbenzoyl)phosphine oxide was used. When the oligomers are irradiated with UV light in the presence of the photoinitiator, two radical-mediated processes occur (102):

1. Thiol-ene coupling, which represents a linear chain extension event, and
2. Acrylamide homopolymerization, which represents the crosslinking mechanism.

The compositions have a low viscosity before printing and the modulus and tensile strain at break of a photo-cured, higher molecular weight precursor after printing. Photorheology and Soxhlet extraction demonstrated highly efficient photocuring, revealing a calculated molecular weight between crosslinks of 12,600 $g\,mol^{-1}$ and gel fractions in excess of 90% while employing significantly lower molecular weight precursors (i.e., smaller than 5300 $g\,mol^{-1}$).

These photo-cured objects demonstrated a twofold increase in tensile strain at break as compared to a photo-cured 5300 $g\,mol^{-1}$ poly(dimethyl siloxane) diacrylamide alone. These results are broadly applicable to the advanced manufacturing of objects requiring high elongation at break (102).

1.8 Hot Lithography

Vat photopolymerization is used for printing very precise and accurate parts from photopolymer resins (106). Conventional 3D printers based on vat photopolymerization are curing resins with low viscosity at or slightly above room temperature. The newly developed hot lithography provides a vat photopolymerization technique, where the resin is heated and cured at elevated temperatures.

The influence of the printing temperature of 23°C and 70°C, on the properties of a printed dimethacrylate resin was shown (106). The working curve was measured for 23°C, 50°C and 70°C. Specimens were printed in both XYZ and ZXY orientation.

The resulting tensile properties were tested, dynamic mechanical analysis was carried out and the double-bond conversion was analyzed. It was found that the critical energy E_0 was significantly reduced by a higher printing temperature. Therefore, the exposure time was reduced from 50 s to 30 s to reach a similar curing depth. A higher printing temperature provided higher double-bond conversion, tensile strength and modulus of the green parts. However, the printing temperature did not affect the properties after post-curing in XYZ orientation. Post-cured tensile specimens in ZXY orientation showed a higher tensile strength when printed at 23°C, because higher overpolymerization led to a smoother surface of the specimens. Overall, higher printing temperatures lowered the viscosity of the resin, reduced the printing time and provided better mechanical properties of green parts while post-cured properties were mostly unaffected (106).

1.9 Ambient Reactive Extrusion

Additive manufacturing has the potential to offer many benefits over traditional manufacturing methods in the fabrication of complex parts (107). The advantages of this technique are low weight, complex geometry, and embedded functionality.

In practice, common additive manufacturing technologies are limited by their slow speed and highly directional properties. A reactive mixture deposition approach has been developed that can enable 3D printing of polymer materials at more than 100 times the volumetric deposition rate, enabled by a greater than tenfold reduction in printhead mass compared to conventional large-scale thermoplastic deposition methods, with material chemistries that can be tuned for specific properties.

In addition, the reaction kinetics and transient rheological properties could be specifically designed for the target deposition rates, so enabling the synchronized development of increasing shear modulus and extensive crosslinking across the printed layers.

This ambient cure eliminates the internal stresses and bulk distortions that typically hamper additive manufacturing of large parts and yields a printed part with interlayer covalent bonds that significantly improve the strength of the part along the build direction.

The fast curing kinetics combined with the fine-tuned viscoelastic properties of the mixture enable the rapid vertical builds that are not possible with other techniques.

In contrast to conventional additive manufacturing approaches, which require larger and slower print systems and complex thermal management strategies as scale increases, liquid reactive polymers decouple performance and print speed from the scale of the part, thus enabling a new class of cost-effective, fuel-efficient additive manufacturing (107).

1.10 Micromanufacturing Engineering

There are monographs dealing with micromanufacturing engineering (108, 109).

1.11 Analytical Uses

1.11.1 Gas Sensors

A self-healing conductive polymer composite ink was formulated for 3D extrusion printing of flexible electronics.

The material was composed of a poly(borosiloxane) matrix and 5% by volume of electrochemically exfoliated graphene (110). The printability was derived from the chemical-activated mechanically adaptive properties of poly(borosiloxane).

The mechanically adaptive properties of poly(borosiloxane) and the composite ink were studied through rheological measurements, and the non-Newtonian nature was analyzed using the Carreau-Yasuda model.

With methanol vapor as a representative stimulus, the underpinning mechanism of the mechanically adaptive properties of poly-(borosiloxane), which involved the methanol-induced alcoholysis of the crosslinking boron/oxygen dative bonds in poly(borosiloxane), was further investigated by infrared spectroscopy. The self-healing, adaptive and conductive composite ink could be used for printing 3D structures and devices on a poly(dimethyl siloxane) flexible substrate. A 3D printed gas sensor with responses to various chemical vapors was demonstrated as a potential application of this composite ink.

1.12 Chemical Engineering

Computer-aided fabrication technologies combined with simulation and data processing approaches are changing the way of manufacturing and designing functional objects. Also, in the field of catalytic technology and chemical engineering the impact of 3D printing is steadily increasing thanks to a rapidly decreasing equipment threshold.

A review has been presented that highlights the research using 3D printing and computational modeling as digital tools for the design and fabrication of reactors and structured catalysts (111).

Computational fluid dynamics is a branch of fluid mechanics that uses numerical analysis to solve fluid flow problems. While computational fluid dynamics was initially driven by aerodynamics, it is now widely applied to geophysical flows in the automotive industry, and to industrial manufacturing (112).

In chemical engineering, computational fluid dynamics involves the numerical solution of conservation equations for mass, momentum and energy in a flow geometry of interest. An application example which has been presented is the study of flow hydrodynamics at different flow rates in three different reactors (113). Computational fluid dynamics allows the quantification of the residence time distribution of each reactor in addition to the description of fluid flow.

In contrast with other manufacturing methods, 3D printing enables the fabrication of geometrically optimized reactors. The shape of conventional reactors is to a large extent determined by their manufacturing cost and/or the limitations of conventional manufacturing methods, e.g., cylindrical shape.

Commonly available reactor shapes may not fully realize the potential of continuous-flow processes. The use of small-scale continuous-flow reactors, with channel cross sections ranging from tens of microns to a few millimeters, increased rapidly over the past two decades. Because of the fast mixing and efficient heat and mass transfer in such flow reactors, a high level of control over reaction parameters is possible (114).

Fully or partially 3D printed continuous-flow reactors will likely have a more profound impact because of their scalability up to industrial production via parallelization. For example, a plug-and-play platform to study the continuous-flow synthesis has been developed, as well as a formulation of active pharmaceutical ingredients (115).

Recently, flow reactors were fabricated using FDM and different polymers, poly(lactic acid) (PLA), high impact poly(styrene), and Nylon (116). The 3D printed flow reactors were tested for a catalytic stereoselective Henry reaction. The Henry reaction, also referred to as the nitro-aldol reaction, is a classic carbon-carbon bond formation reaction in organic chemistry (117). Biologically active targets, such as norephedrine, metaraminol, and methoxamine, were synthesized in a two-step continuous-flow process inside the customizable reactors.

Also, 3D printing allows the direct integration of active flow control components such as valves before, after or inside the main reactor chamber. Especially at the small scale it could be beneficial to include such control operators in a fully automated and integrated microreactor.

In addition to 3D printed structured catalysts being considered excellent research tools, they might also find application in the commercial production of small to medium catalyst batches with custom geometries or sizes. Similarly, robocast ceramic lattice filters have found application as customizable filtration media for melted metal in foundries (118). Further opportunities are likely to be found in application areas closely related to structured catalysts such as structured column internals for reactive distillations, 3D printed chromatographic media, and optimized heat exchanger internals (111).

1.12.1 *Gas Separation*

Organic-inorganic materials have emerged as promising candidates for use in various gas separation processes due to the beneficial synergistic effect of both constituents.

The formulation of various adsorbents, including zeolite 13X, zeolite 5A, silica-supported amines, MOF-74(Ni), and UTSA-16(Co) into monolithic contactors with adsorbent loadings as high as 90%, using 3D printing technique has been demonstrated (119–121).

Also, a facile approach for the fabrication of composite monoliths was reported consisting of Torlon polymer and 13X and 5A zeolites via 3D printing technique for the removal of CO_2 from flue gas (122).

The composition of the final dope is presented in Table 1.7.

Table 1.7 Compositions (122).

Monolith	Torlon	Zeolite	PVP /[%]	NMP	DI water
Torlon	25	–	7	63	5
Torlon-zeolite 13X	18	13	3	59	7
Torlon-zeolite 5A	18	13	3	59	7

The physical and structural properties of the 3D printed composite monoliths were systematically evaluated and compared with their pristine powders.

The formation of the monoliths was facilitated through a non-solvent-induced phase separation technique. By the incorporation of around 31% zeolite particles into the polymer matrix, the obtained monoliths displayed CO_2 capture capacities proportional to the zeolite loading.

The CO_2 adsorption performance of the adsorbents was evaluated and compared with their corresponding powders. The 3D printed monolithic adsorbents showed a comparable CO_2 capacity, fast adsorption kinetics, relative stability, and regenerability to their powder counterparts.

The monoliths displayed a similar isotherm to their powder analogues with a steep uptake at low pressures, up to 0.15 *bar*, indicative of rapid micropore filling as generally observed in microporous materials, followed by a gradual increase in CO_2 uptake at higher pressures, from 0.15 to 1 *bar*. The measured CO_2 uptakes of 3D printed T-13X-M and T-5A-M monoliths were 1.83 and 1.51 $mmol\,g^{-1}$, respectively, in comparison to 4.04 and 3.64 $mmol\,g^{-1}$ obtained for their powder analogues at 35°C and 1 *bar*.

Both 3D printed Torlon-zeolite monoliths exhibited high compressive strengths of approximately 210 *MPa*, which was significantly higher than that of 3D printed zeolite monoliths developed earlier (119–121).

These results demonstrate the superiority of polymer-zeolite composite monoliths formulated by 3D printing technique as robust structures with outstanding mechanical integrity and comparable adsorption capacity for various adsorption-based gas separation processes (122).

1.12.2 Hierarchical Monoliths for Carbon Monoxide Methanation

In general, monoliths exhibit better heat and mass transfer properties and are thus potentially favorable in situations such as highly exothermal reactions (123). The key advantage of 3D printed catalysts as compared to that of the conventionally prepared catalysts is the design of tortuous channels that modulate the transport properties in hierarchical monoliths rather than that of simple, straight or uncontrollable channels.

Cylinder, tetrahedron, and tetrakaidecahedron periodic structures were modulated via 3D printing. They could be used as a hard template to prepare phenol-formaldehyde-based hierarchical monoliths for CO methanation. The reaction results showed that a 3D monolith with a 1 *mm* diameter simple straight channel has a high catalytic performance at similar loadings and compositions of active components. The channel structure can be precisely modulated by 3D printing, and the macro-channels and meso-channels are well connected.

Through modulating the tortuosity of the macro-channels, the $Ni\text{-}Al_2O_3/C$ monolith with a 1.25 *mm* diameter tetrahedral channel showed an excellent CH_4 yield as well as a prominent decrease in the temperature gradient and a pressure at 24,000 h^{-1} gas hourly space velocity, in comparison to other structures of monolithic catalysts.

A high mass and a high heat transport along with reaction activity efficiently facilitate the improvement of the performance of the catalyst (123).

1.13 Rotating Spinnerets

The most common process for fabricating nanofibers is electrospinning (124). Electrospinning is a process that uses high voltage to create an electric field between a droplet of polymer solution at the tip of a needle and a collector plate. One electrode of the voltage source is placed into the solution and the other is connected to the collector. This creates an electrostatic force. As the voltage is increased the electric field intensifies, causing a force to build up on the pendant drop of polymer solution at the tip of the needle. This force acts in a direction opposing the surface tension of the drop. The increasing electrostatic force causes the drop to elongate, forming a conical shape known as a Taylor cone. When the electrostatic force overcomes the surface tension of the drop, a charged, continuous jet of solution is ejected from the cone. The jet of solution accelerates towards the collector, whipping and bending wildly. As the solution moves away from the needle and toward the collector, the jet rapidly thins and dries as the solvent evaporates. On the surface of the grounded collector, a nonwoven mat of randomly oriented solid nanofibers is deposited.

However, there are multiple drawbacks associated with electrospinning, such as the requirement for a high voltage electrical field, low production rate, the requirement for precise solution conductivity, and the need for additional devices for producing aligned fiber structures (124).

A device for the fabrication of a micron, submicron or nanometer dimension polymeric fiber has been developed. The device includes an oscillating track system, said system comprising a reservoir suitable for accepting a polymer and operably linked to the track system and comprising an orifice for ejecting said polymer during oscillation, e.g., vertical, horizontal, or diagonal oscillation, of the reservoir along the track system, thereby forming a micron, submicron or nanometer dimension polymeric fiber, and a collector for accepting said formed micron, submicron or nanometer dimension polymeric fiber, wherein the device is free of an electrical field, e.g., a high voltage electrical field (124). A schematic representation of the device is shown in Figure 1.5. It shows that the polymer solution is ejected from the two orifices of the rotating reservoir due to centrifugal forces.

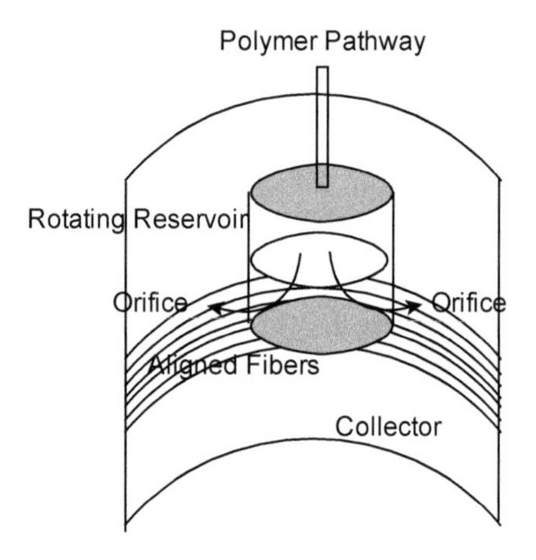

Figure 1.5 Schematic representation of the devices (124).

Sufficient speeds and times for rotating the rotary spinning system or oscillating the oscillating track system to form a polymeric fiber are dependent on the concentration of the polymer and the desired features of the formed polymeric fiber. For example, an 8% weight solution of PLA rotated at 10,000 *rpm* allowed the formation of continuous polymeric fibers.

The methods include mixing a biologically active agent, e.g., a polypeptide, protein, nucleic acid molecule, nucleotide, lipid, biocide, antimicrobial, or pharmaceutically active agent, with the polymer during the fabrication process of the polymeric fibers. For example, polymeric fibers prepared using the device were contacted with encapsulated fluorescent poly(styrene) beads (124). Also, a plurality of living cells can be mixed with the polymer during the fabrication process of the polymeric fibers. Here, biocompatible polymers, e.g., hydrogels, may be used (124).

1.14 Objects with Surface Microstructures

Polymer surface structures, which govern a variety of material properties, such as wetting, adsorption, catalyzing, friction, adhesion, and light absorption/refraction, have become highly attractive. The properties of polymer surface depend on their chemical compositions and surface microstructures as well as their synergy (125). Recent advances in polymer microstructures have led to the emergence of a variety of applications, including super-hydrophobic self-cleaning materials, photonic crystal, and microfluidics (126)

Super-hydrophobic and self-cleaning surfaces have been detailed (127). Super-hydrophobic surfaces having a water contact angle greater than 150° and a water slip-off angle less than 10° can have many potential applications, from small non-wetting micro/nano-electronics to large self-cleaning industrial equipment.

A method for preparing stable 3D polymer objects with surface microstructures or nanostructures has been described. The method includes the following steps (125):

1. Synthesizing a thermoset 2D polymer object with surface microstructures. The polymer network contains reversible exchangeable bonds.
2. Deforming synthesized polymer to an arbitrary desired shape above the reshaping temperature with an application of an external force. The permanent reshaping temperature falls in the range of 50°C–130°C and external stress is held for 5 *min* – 24 *h*.
3. After cooling, a permanent 3D polymer object with surface microstructure is obtained.

Steps 2 and 3 can be repeated for many cycles and the 2D polymer object can be arbitrarily and cumulatively deformed to get a complex 3D structures. The polymer networks contain reversible exchangeable bonds and bond exchange catalysts.

The polymers synthesized in step 1 are a poly(urethane) resin, a poly(urethane)-urea resin, an acid/anhydride cured epoxy resin, silicone, a resin with multiple hydrogen bonds, or Diels-Alder reaction products. Polyisocyanates can be chosen from poly(hexamethylene diisocyanate), triphenylmethane-4,4′,4′′-triisocyanate, and a 1,6-hexamethylene diisocyanate tripolymer.

Preparation 1–2: First 0.2 *mmol* poly(tetrahydrofuran) diol, 0.2 *mmol* poly(hexamethylene diisocyanate) were added to 10 *ml* *N*,*N*-dimethylformamide and melted by heating in an oven at 80°C, wherein mass ratio of poly(caprolactone) and poly(hexamethylene diisocyanate) is the mole ratio of hydroxyl group and isocyanates group. Then a predetermined amount of DBTDL (0.5% of total weight) and 1,5,7-triazabicyclo[4.4.0]dec-5-ene (2% of total weight) were dissolved into the mixture and stirred for several minutes. The mixture was poured into the mold with specific microstructures and curing was conducted thermally at 90°C for 12 *h*. After curing completely, the sample was demolded. Afterwards, the sample was bent into a 3D surface and thermally annealed under stress (130°C, 10 *min*), then cooled. Finally, a 3D surface with micro-nanostructures was obtained.

Also, compositions and methods for preparation of the other above-mentioned systems have been detailed (125).

1.15 Lightweight Cellular Composites

An epoxy-based ink has been described that enables the 3D printing of lightweight cellular composites with a controlled alignment of multiscale, high aspect ratio fiber reinforcement to create hierarchical structures inspired by balsa wood (128). Young's modulus values up to 10 times higher than commercially available 3D printed polymers are attainable, while comparable strength values can be maintained.

To fabricate the lightweight cellular composites, epoxy-based inks are formulated that embody the essential rheological properties required for the 3D printing method.

Epoxy resins are reactive materials that initially exhibit a low viscosity, which rises over time as the reaction proceeds under ambient conditions. Moreover, these inks ultimately require thermal curing at elevated temperatures of 100°C to 220°C for several hours to complete the polymerization process.

The inks are prepared by mixing an epoxy resin (Epon 826 epoxy resin, Momentive Specialty Chemicals, Inc., Columbus, OH) with appropriate amounts of dimethyl methyl phosphonate (DMMP, Sigma-Aldrich, St. Louis, MO), nanoclay platelets (Cloisite 30B, Southern Clay Products, Inc., Gonzales, TX), silicon carbide whiskers (SI-TUFF SC-050, ACM, Greer, SC), and milled carbon fibers (Dialead K223HM, Mitsubishi Plastics, Inc., Tokyo, Japan) using a Thinky

Planetary Centrifugal Mixer (Thinky USA, Inc., Laguna Hills, CA) in a 125 *ml* glass container using a custom adaptor. An imidazole-based ionic liquid 1-ethyl-3-methylimidazolium dicyanamide is used as latent curing agent (Basionics VS 03, BASF Intermediates, Ludwigshafen, Germany) (128).

1.16 Textiles

1.16.1 3D Printed Polymers Combined with Textiles

Composites that have combined two or more different materials with different physical and chemical properties allow the tailoring of mechanical and other characteristics of the resulting multimaterial system (129).

In relation to fiber reinforced plastic composites, combinations of textile materials with 3D printed polymers result in different mechanical properties. While the tensile strength of the multimaterial system is increased in comparison to the pure 3D printed material, the elasticity of the polymer layer can be retained to a certain degree, since the textile material is not completely immersed in the polymer. Instead, an interface layer is built in which both materials interpenetrate to a certain degree.

The adhesion between both materials at this interface has been investigated (129). It was shown that both the printing material and the textile substrate influence the adhesion between both materials due to viscosity during printing, thickness and pore sizes, respectively.

Some material combinations build strong form-locking connections, other compositions can easily be delaminated. Depending on both materials, significantly different adhesion values were found in such 3D printed composites.

These properties make some combinations very well suitable for building composites with novel mechanical properties, while others suffer from insufficient connections (129).

1.16.2 Mechanical and Electrical Contacting

Nowadays materials are becoming smarter and more intelligent, and contain novel functionalities or sensory components (130).

In the area of textiles, several functions can be achieved by finishing techniques, i.e., physical and/or chemical modifications of the surfaces of textiles. The integration of electronic components, however, may still suffer from incompatibilities between soft, flexible, bendable textile fabrics and rigid electronic parts. Connecting conductive yarns with electronic components, such as SMD-LEDs, etc., cannot be performed by soldering or sewing. The typical connection technologies of both areas fail in these cases.

A new possibility to achieve such electrical and at the same time mechanical connections is given by 3D printing. The chances and limitations of electric circuits combining textile fabrics with 3D printing have been studied (130).

In textile fabrics woven and knitted from common yarns as well as wires, strands and different conductive yarns, a 3D printing technique was used to connect SMD elements and other small electronic parts with these base circuits and compared with soldered and sewn contacts. The influence of conductive wires, yarns or filaments integrated in the printed elements was tested.

The possibilities and problems in electrical contacting of small electronic components on partly conductive textile fabrics by conductive 3D printed connections have been detailed. Also, an outlook on other potential areas of application, such as sensors and actuators on textile fabrics, has been given (130).

1.16.3 *Soft Electronic Textiles*

E-textiles (electronic textiles) are fabrics that possess electronic counterparts and electrical interconnects knitted into them, offering flexibility, stretchability, and a characteristic length scale that cannot be accomplished using other electronic manufacturing methods currently available (131). The recent advances in the field of e-textiles and particularly the materials and their functionalities have been detailed (131).

Potential applications of e-textiles are collected in Table 1.8.

In e-textiles, the components and the interconnections are barely visible, since they are connected intrinsically to soft fabrics that have attracted the attention of those in the fashion and textile industries. These textiles can effortlessly acclimatize themselves to

Table 1.8 Potential applications of e-textiles (131).

Applications	Literature
Proximity sensors	(132)
Heat control	(133)
Humidity	(134)
Optical guides	(135)
Realization of light diffusers	(136)
Biophysical sensing abilities	(137–140)

the fast-changing wearable electronic markets with digital, computational, energy storage, and sensing requirements of any specific application.

Knitting is only one of the technologies in e-textile integration. Other technologies, such as sewing, embroidery, and even single fiber-based manufacture technology, are widely employed in next-generation e-textiles.

Electrospinning and melt spinning techniques have been used to produce polymeric nanofibers and microfibers. These materials have thereafter been carbonized to fabricate conducting carbon fabrics.

In some cases, electrospinning was performed on textiles to produce wearable electronics. Electrospun carbon nanowebs on metallic textiles were used to produce high-capacitance supercapacitor fabrics (141).

Also, a method of integrating a substrate-less electrospinning process with textile technology has been described (142). Here, a new collector design was developed that provides a pressure-driven, localized cotton-wool structure in free space from which continuous high-strength yarns were drawn. An advantage of this integration was that the textile could be drug/dye loaded and be developed into a core-sheath architecture with a greater functionality. This method could produce potential nano-textiles for various biomedical applications.

The fabrication of conductive, flexible, and durable textiles with reduced graphene oxide on nylon-6 yarns, cotton yarns, polyester yarns, and nonwoven textile fabrics via a facile preparation method has been reported (143). Here, the problem of adhesion between graphene oxide and textiles could also be solved by the use of bovine serum albumin proteins, which serve as universal adhesives for any textile, regardless of its surface condition.

Environmental parameter sensors, such as heat, humidity, and pressure sensors, have been attached in textiles (144).

Microfluidics and 3D printing can be combined in fabric-based point-of-care diagnostic applications (130, 145).

Integrated with light-emitting diodes, these fabrics can be used for visual sensing (146). The programmable delivery and release of therapeutic drugs is also possible by combining drug-loaded micro/nanoparticles with e-textiles (147, 148).

An actively controlled wound dressing has been developed using composite fibers attached to a heating element and enclosed by a hydrogel layer comprised of thermoresponsive drugs (149). The drug- and biological-factor-loaded fibers were assembled using textiles and a flexible and wearable wound dressing was created. Such fibers can individually address the programmable release of different drugs with a collective temporal profile.

An efficient monitoring method of sweat pH was reported by a wearable sensor based on a cotton fabric treated with an organically modified silicate and miniaturized and low-power electronics with wireless interfaces (150).

Furthermore, a fully textile, wearable organic electrochemical transistor sensor for the detection of biomarkers in external body fluids was developed without using an invasive electrode (151).

1.16.4 4D Textiles

Additive manufacturing combined with highly elastic, extensible textile materials provides the opportunity to explore a new range of materials: 4D textiles (152). In the case of 4D textiles, the time response is necessary, but the textile material also plays a crucial role in responding to external stimuli.

Whereas 4D printing is currently limited to very small deformations and very slow changes in time, 4D textiles offer the opportunity to increase the deformation and the response time.

The concepts of 4D printing, achievements in 3D printing, and the concept of 4D textiles have been reported. Also, the role of materials, critical process parameters, critical textile processes, and potential application areas have been presented (152).

Stimuli that can activate over time are: Sound, waves, humidity, UV, chemical, temperature, mechanical, and electrical.

Textile materials can be integrated into 3D printing with various forms and functions. As functional fibers they are integrated into composites that can be used for shape memory effects.

References

1. Wikipedia contributors, 3d printing — Wikipedia, the free encyclopedia, https://en.wikipedia.org/w/index.php?title=3D_printing&oldid=837253889, 2018. [Online; accessed 20-April-2018].
2. K. Roebuck, *3D Printing*, Emereo Pty. Ltd, 2011.
3. C. Barnatt, *3D printing: The next industrial revolution*, ExplainingThe-Future.com, S.l, 2013.
4. H. Lipson, *Fabricated: The New World of 3D Printing*, John Wiley & Sons, Indianapolis, Indiana, 2013.
5. T. Neill, *3D Printing*, Cherry Lake Publishing, Ann Arbor, Michigan, 2013.
6. C. Thorpe, *Adventures in 3D Printing and Design*, John Wiley & Sons Inc., Hoboken (New Jersey), 2015.
7. F. Bitonti, *3D Printing for Fashion: An Introduction to Coding Couture*, Fairchild Books, London, UK New York, NY, USA, 2017.
8. C. Chua, *Standards, Quality Control, and Measurement Sciences in 3D Printing and Additive Manufacturing*, Academic Press, an imprint of Elsevier, London, UK, 2017.
9. P. Kocovic, *3D Printing and Its Impact on the Production of Fully Functional Components: Emerging Research and Opportunities*, Engineering Science Reference, Hershey, PA, 2017.
10. S. Magdassi, *Nanomaterials for 2D and 3D Printing*, Wiley-VCH, Weinheim, Germany, 2017.
11. R. Noorani, *3D Printing*, CRC Press, London, 2017.
12. A. Ovsianikov, *3D Printing and Biofabrication*, Springer International PU, Cham, 2017.
13. F. Rybicki, *3D Printing in Medicine: A Practical Guide for Medical Professionals*, Springer, Cham, Switzerland, 2017.
14. S. Hoskins, *3D Printing for Artists, Designers and Makers*, Bloomsbury Visual Arts, An imprint of Bloomsbury Publishing PLC, London New York, 2018.
15. A. McMills, *3D Printing Basics for Entertainment Design*, Routledge, New York, NY, 2018.

16. J.M.B. Mosadegh, S.D. Subhi, and J. Al'Aref, eds., *3D Printing Applications in Cardiovascular Medicine*, Elsevier Academic Press, Cambridge, MA, 2018.
17. J. Perritano, *3D Printing*, Saddleback Educational Publishing, Costa Mesa, CA, 2018.
18. S. Torta, *3D Printing: An Introduction*, Mercury Learning & Information, Herndon, VA, 2018.
19. R. Cameron, *Mastering 3D Printing in the Classroom, Library, and Lab*, Apress, Cambridge, MA, 2019.
20. D. Dimitrov, N. de Beer, P. Hugo, and K. Schreve, Three dimensional printing in S. Hashmi, G.F. Batalha, C.J.V. Tyne, and B. Yilbas, eds., *Comprehensive Materials Processing*, Vol. 10, pp. 217–250. Elsevier, Oxford, 2014.
21. M.J. Cima, E. Sachs, L.G. Cima, J. Yoo, S. Khanuja, S.W. Borland, B. Wu, and R.A. Giordano, Computer-derived microstructures by 3D printing: Bio- and structural materials, in *Solid Freeform Fabr Symp Proc: DTIC Document*, pp. 181–90, 1994.
22. B.M. Wu, S.W. Borland, R.A. Giordano, L.G. Cima, E.M. Sachs, and M.J. Cima, *Journal of Controlled Release*, Vol. 40, p. 77, 1996.
23. T. Billiet, M. Vandenhaute, J. Schelfhout, S.V. Vlierberghe, and P. Dubruel, *Biomaterials*, Vol. 33, p. 6020, 2012.
24. B.M. Wu and M.J. Cima, *Polymer Engineering & Science*, Vol. 39, p. 249, 2004.
25. H.N. Chia and B.M. Wu, *Journal of Biological Engineering*, Vol. 9, p. 4, 2015.
26. J. Horvath, A brief history of 3D printing in *Mastering 3D Printing,*, pp. 3–10. Apress, Berkeley, CA, 2014.
27. M. Whitaker, *The Bulletin of the Royal College of Surgeons of England*, Vol. 96, p. 228, 2014.
28. Wikipedia, Chuck Hull — wikipedia, the free encyclopedia, 2014. [Online; accessed 30-July-2014].
29. C.W. Hull, Apparatus for production of three-dimensional objects by stereolithography, US Patent 4 575 330, assigned to UVP, Inc. (San Gabriel, CA), March 11, 1986.
30. B. Evans, *Practical 3D Printers*, Apress, New York, 2012.
31. H. Kodama, *Review of Scientific Instruments*, Vol. 52, p. 1770, 1981.
32. AV Plastics, 3D printing history, http://www.avplastics.co.uk/3d-printing-history, 2018.
33. P. Parandoush and D. Lin, *Composite Structures*, Vol. 182, p. 36 , 2017.
34. A. Pye, 3D printing - a convergence between printer makers and polymer manufacturers, Technical review, UL Prospector, Santa Fe, KS, 2018. electronic: https://knowledge.ulprospector.com/8181/.
35. M. Layani, X. Wang, and S. Magdassi, *Advanced Materials*, Vol. 0, p. 1706344, 2018.

36. G. Bugeda, M. Cervera, G. Lombera, and E. Onate, *Rapid Prototyping Journal*, Vol. 1, p. 13, 1995.
37. Y. Tang, C. Henderson, J. Muzzy, and D.W. Rosen, *International Journal of Materials and Product Technology*, Vol. 21, p. 255, 2004.
38. Q. Huang, 3D printing shrinkage compensation using radial and angular layer perimeter point information, US Patent 9 886 526, assigned to University of Southern California (Los Angeles, CA), February 6, 2018.
39. J.G. Zhou, D. Herscovici, and C.C. Chen, *International Journal of Machine Tools and Manufacture*, Vol. 40, p. 363, 2000.
40. C. Lynn-Charney and D.W. Rosen, *Rapid Prototyping Journal*, Vol. 6, p. 77, 2000.
41. S.O. Onuh and K.K.B. Hon, *The International Journal of Advanced Manufacturing Technology*, Vol. 17, p. 61, Jan 2001.
42. K. Tong, E.A. Lehtihet, and S. Joshi, *Rapid Prototyping Journal*, Vol. 9, p. 301, 2003.
43. K. Tong, S. Joshi, and E.A. Lehtihet, *Rapid Prototyping Journal*, Vol. 14, p. 4, 2008.
44. X. Wang, *Rapid Prototyping Journal*, Vol. 5, p. 129, 1999.
45. P.D. Hilton and P.F. Jacobs, *Rapid Tooling: Technologies and Industrial Applications*, Marcel Dekker, New York, 2000.
46. B. Storåkers, N.A. Fleck, and R.M. McMeeking, *Journal of the Mechanics and Physics of Solids*, Vol. 47, p. 785 , 1999.
47. J. Secondi, *Powder Metallurgy*, Vol. 45, p. 213, 2002.
48. K. Mori, K. Osakada, and S. Takaoka, *Engineering Computations*, Vol. 13, p. 111, 1996.
49. D.L. Bourell, M.C. Leu, and D.W. Rosen, Roadmap for additive manufacturing: Identifying the future of freeform processing, Laboratory for freeform fabrication advanced manufacturing center, The University of Texas at Austin, Austin, TX, 2009.
50. D. Russell, A. Hernandez, J. Kinsley, and A. Berlin, Apparatus and methods for 3D printing, US Patent 7 291 002, assigned to Z Corporation (Burlington, MA), November 6, 2007.
51. J.F. Bredt, T.C. Anderson, and D.B. Russell, Three dimensional printing materials system, US Patent 6 416 850, assigned to Z Corporation (Burlington, MA), July 9, 2002.
52. Y. Zhou, W.M. Huang, S.F. Kang, X.L. Wu, H.B. Lu, J. Fu, and H. Cui, *Journal of Mechanical Science and Technology*, Vol. 29, p. 4281, Oct 2015.
53. D.C. Deisenroth, R. Moradi, A.H. Shooshtari, F. Singer, A. Bar-Cohen, and M. Ohadi, *Heat Transfer Engineering*, Vol. 0, p. 1, 2017.
54. C.D. Weber, C.G. Dupuy, J.P. Harmon, and D.M. Schut, Inks for 3D printing gradient refractive index (GRIN) optical components, US Patent 9 771 490, assigned to Vadient Optics, LLC (Beaverton, OR), September 26, 2017.

55. X. Xu and M.E. Savard, GRIN lens and method of manufacturing, US Patent 5 689 374, assigned to LightPath Technologies, Inc. (Albuquerque, NM), November 18, 1997.
56. R. Chartoff, B. McMorrow, and P. Lucas, Functionally graded polymer matrix nano-composites by solid freeform fabrication, in D.L. Bourell, R.H. Crawford, J.J. Beaman, K.L. Wood, and H.L. Marcus, eds., *Solid Freeform Fabrication Proceedings*, pp. 365–391, Austin, TX., 2003. University of Texas at Austin.
57. B. McMorrow, R. Chartoff, P. Lucas, and W. Richardson, Polymer matrix nanocomposites by inkjet printing, in D.L. Bourell, R.H. Crawford, J.J. Beaman, K.L. Wood, and H.L. Marcus, eds., *Solid Freeform Fabrication Proceedings*, pp. 174–183, Austin, TX, 2005. University of Texas at Austin.
58. J.A. Lawton and J.T. Adams, Method for fabricating an integral three-dimensional object from layers of a photoformable composition, US Patent 5 122 441, assigned to E. I. Du Pont de Nemours and Company (Wilmington, DE), June 16, 1992.
59. G.H. Kirby, D.J. Harris, Q. Li, and J.A. Lewis, *Journal of the American Ceramic Society*, Vol. 87, p. 181, February 2004.
60. J.A. Lewis, G. Kirby, J.H.-W. Cheung, and A.A. Jeknavorian, Controlled dispersion of colloidal suspension by comb polymers, US Patent 7 053 125, assigned to The Board of Trustees of the University of Illinois (Urbana, IL) W.R. Grace & Co.-Conn. (Columbia, MD), May 30, 2006.
61. Q. Li and J. Lewis, *Advanced Materials*, Vol. 15, p. 1639, October 2003.
62. J.A. Lewis, Q. Li, and R. Rao, Biphasic inks, US Patent 8 187 500, assigned to The Board of Trustees of the University of Illinois (Urbana, IL), May 29, 2012.
63. E. Giller and D.X. Williams, Three dimensional printing material system and method using plasticizer-assisted sintering, US Patent 8 506 862, assigned to 3D Systems, Inc. (Rock Hill, SC), August 13, 2013.
64. C.L.M. Palenzuela and M. Pumera, *TrAC Trends in Analytical Chemistry*, Vol. 103, p. 110, 2018.
65. A. Moien, G. Albert, T. Moritz, S. Minseok, P. Elisabeth, and K. Eugenia, *Advanced Materials Technologies*, Vol. 3, p. 1800068, 2018.
66. S.A.N. Gowers, V.F. Curto, C.A. Seneci, C. Wang, S. Anastasova, P. Vadgama, G.-Z. Yang, and M.G. Boutelle, *Analytical Chemistry*, Vol. 87, p. 7763, 2015.
67. W. Lee, D. Kwon, W. Choi, G.Y. Jung, A.K. Au, A. Folch, and S. Jeon, *Scientific Reports*, Vol. 5, p. 7717, 2015.
68. P. Huang, Z. Xia, and S. Cui, *Materials & Design*, Vol. 142, p. 11, 2018.
69. K. Kadimisetty, S. Malla, and J.F. Rusling, *ACS Sensors*, Vol. 2, p. 670, 2017.

70. L. Dennany, R.J. Forster, and J.F. Rusling, *Journal of the American Chemical Society*, Vol. 125, p. 5213, 2003.
71. C. Park, J. Lee, Y. Kim, J. Kim, J. Lee, and S. Park, *Journal of Microbiological Methods*, Vol. 132, p. 128, 2017.
72. I. Ganesh, B.M. Tran, Y. Kim, J. Kim, H. Cheng, N.Y. Lee, and S. Park, *Biomedical Microdevices*, Vol. 18, p. 116, Dec 2016.
73. M.F. Santangelo, S. Libertino, A.P.F. Turner, D. Filippini, and W.C. Mak, *Biosensors and Bioelectronics*, Vol. 99, p. 464, 2018.
74. J.F. Bredt, S.L. Clark, D.X. Williams, and M.J. DiCologero, Thermoplastic powder material system for appearance models from 3D printing systems, US Patent 7 569 273, assigned to Z Corporation (Burlington, MA), August 4, 2009.
75. W.F. Jandeska, Jr. and J.E. Hetzner, Aluminum/magnesium 3D-printing rapid prototyping, US Patent 7 141 207, assigned to General Motors Corporation (Detroit, MI), November 28, 2006.
76. P. Bartolo and G. Mitchell, *Rapid Prototyping Journal*, Vol. 9, p. 150, 2003.
77. J.A. Lewis, J.E. Smay, J. Stuecker, and J. Cesarano, *Journal of the American Ceramic Society*, Vol. 89, p. 3599, 2006.
78. K. Salonitis, *Comprehensive Materials Processing*, Vol. 10, p. 19, 2014.
79. D.L. Bourell, H.L. Marcus, J.W. Barlow, J.J. Beaman, and C.R. Deckard, Multiple material systems for selective beam sintering, US Patent 5 076 869, assigned to Board of Regents, The University of Texas System (Austin, TX), December 31, 1991.
80. R. Vilar, *Comprehensive Materials Processing*, Vol. 10, p. 163, 2014.
81. V.M. Weerasinghe and W.M. Steen, Laser cladding with pneumatic powder delivery in O.D.D. Soares and M. Perez-Amor, eds., *Applied Laser Tooling*, pp. 183–211. Springer Netherlands, 1987.
82. E.M. Sachs, J.S. Haggerty, M.J. Cima, and P.A. Williams, Three-dimensional printing techniques, US Patent 5 204 055, assigned to Massachusetts Institute of Technology (Cambridge, MA), April 20, 1993.
83. M. Cima, E. Sachs, T. Fan, J.F. Bredt, S.P. Michaels, S. Khanuja, A. Lauder, S.-J.J. Lee, D. Brancazio, A. Curodeau, and H. Tuerck, Three-dimensional printing techniques, US Patent 5 387 380, assigned to Massachusetts Institute of Technology, February 7, 1995.
84. E.M. Sachs, Powder dispensing apparatus using vibration, US Patent 6 036 777, assigned to Massachusetts Institute of Technology (Cambridge, MA), March 14, 2000.
85. J.R. Anderson, D.T. Chiu, R.J. Jackman, O. Cherniavskaya, J.C. McDonald, H. Wu, S.H. Whitesides, and G.M. Whitesides, *Analytical Chemistry*, Vol. 72, p. 3158, July 2000.

86. J. Liu and M. Rynerson, Method for article fabrication using carbohydrate binder, US Patent 6 585 930, assigned to Extrude Hone Corporation (Irwin, PA), July 1, 2003.
87. B.G. Compton and J.A. Lewis, *Advanced Materials*, Vol. 26, p. 5930, June 2014.
88. R. Rael, 3D printing powder compositions and methods of use, WO Patent 2 013 043 908, assigned to The Regents of the University of California, March 28, 2013.
89. T. Ito, T. Hagiwara, T. Ozai, and T. Miyao, Rapid prototyping resin compositions, US Patent 8 293 810, assigned to CMET Inc. (Yokohama-Shi, JP) Shin-Etsu Chemical Co., Ltd. (Tokyo, JP), October 23, 2012.
90. J. Southwell, B.A. Register, S.K. Sarmah, P.A.M. Steeman, B.J. Keestra, and M.M. Driessen, Radiation curable resin composition and rapid three-dimensional imaging process using the same, US Patent 8 501 033, assigned to DSM IP Assets B.V. (Heerlen, NL), August 6, 2013.
91. N.D. Dolinski, Z.A. Page, E.B. Callaway, F. Eisenreich, R.V. Garcia, R. Chavez, D.P. Bothman, S. Hecht, F.W. Zok, and C.J. Hawker, *Advanced Materials*, Vol. 0, p. 1800364, 2018.
92. G. Terrones and A.J. Pearlstein, *Macromolecules*, Vol. 34, p. 3195, 2001.
93. G.A. Miller, L. Gou, V. Narayanan, and A.B. Scranton, *Journal of Polymer Science Part A: Polymer Chemistry*, Vol. 40, p. 793, 2002.
94. N.D. Dolinski, Z.A. Page, F. Eisenreich, J. Niu, S. Hecht, J. Read de Alaniz, and C.J. Hawker, *ChemPhotoChem*, Vol. 1, p. 125, 2017.
95. T.F. Patrício, R.F. Pereira, A. Cerva, and P.J. Bártolo, VAT polymerization techniques for biotechnology and medicine, in *High Value Manufacturing: Advanced Research in Virtual and Rapid Prototyping: Proceedings of the 6th International Conference on Advanced Research in Virtual and Rapid Prototyping, Leiria, Portugal, 1-5 October, 2013*, p. 203. CRC Press, 2013.
96. I. Gibson, D. Rosen, and B. Stucker, Vat photopolymerization processes, in *Additive Manufacturing Technologies: 3D Printing, Rapid Prototyping, and Direct Digital Manufacturing*, pp. 63–106, New York, NY, 2015. Springer New York.
97. M. Mischkot, T. Hofstätter, N. Bey, D.B. Pedersen, H.N. Hansen, and M.Z. Hauschild, Life cycle assessment injection mold inserts: Additively manufactured, in brass, and in steel, in *Proceedings of the DTU Sustain Conference*, Lyngby, Denmark, 2015. DTU Library, Technical information Center of Denmark.
98. T. Hofstätter, N. Bey, M. Mischkot, A. Lunzer, D.B. Pedersen, and H.N. Hansen, Comparison of conventional injection mould inserts to additively manufactured inserts using life cycle assessment, in *EUSPEN 16th International Conference & Exhibition*, Nottingham, UK,

2016. European Society for Precision Engineering and Nanotechnology.
99. T. Hofstätter, D.B. Pedersen, G. Tosello, and H.N. Hansen, *Procedia CIRP*, Vol. 66, p. 312 , 2017. 1st CIRP Conference on Composite Materials Parts Manufacturing (CIRP CCMPM 2017).
100. Fun To Do, Affordable 3D resins, technical data sheet, electronic: http://www.funtodo.net/properties.html, 2018.
101. Additive Manufacturing Research Group, About additive manufacturing, vat photopolymerisation, electronic: http://www.lboro.ac.uk/research/amrg/about/the7categoriesofadditivemanufacturing/vatphotopolymerisation/, 2018. Loughborough University, Leicestershire, UK.
102. J.M. Sirrine, V. Meenakshisundaram, N.G. Moon, P.J. Scott, R.J. Mondschein, T.F. Weiseman, C.B. Williams, and T.E. Long, *Polymer*, 2018. in press.
103. L. Kürti and B. Czakó, *Strategic Applications of Named Reactions in Organic Synthesis: Background and Detailed Mechanisms*, Elsevier Science, Burlington, 2005.
104. Wikipedia contributors, Schotten-Baumann reaction — Wikipedia, the free encyclopedia, https://en.wikipedia.org/w/index.php?title=Schotten%E2%80%93Baumann_reaction&oldid=850522116, 2018. [Online; accessed 23-July-2018].
105. Wikipedia contributors, Fischer-Speier esterification — Wikipedia, the free encyclopedia, https://en.wikipedia.org/w/index.php?title=Fischer%E2%80%93Speier_esterification&oldid=850394315, 2018. [Online; accessed 23-July-2018].
106. B. Steyrer, B. Busetti, G. Harakály, R. Liska, and J. Stampfl, *Additive Manufacturing*, Vol. 21, p. 209, 2018.
107. O. Rios, W. Carter, B. Post, P. Lloyd, D. Fenn, C. Kutchko, R. Rock, K. Olson, and B. Compton, *Materials Today Communications*, Vol. 15, p. 333, 2018.
108. Y. Qin, ed., *Micromanufacturing Engineering and Technology*, Micro and Nano Technologies, William Andrew Publishing, Boston, 2nd edition, 2015.
109. I. Choudhury, *Finish Machining and Net-Shape Forming*, Elsevier, Waltham MA, 2017.
110. T. Wu, E. Gray, and B. Chen, *J. Mater. Chem. C*, Vol. 6, p. 6200, 2018.
111. C. Parra-Cabrera, C. Achille, S. Kuhn, and R. Ameloot, *Chemical Society Reviews*, Vol. 47, p. 209, 2018.
112. J.A.M. Kuipers and W.P.M. van Swaaij, Computational fluid dynamics applied to chemical reaction engineeringVol. 24 of *Advances in Chemical Engineering*, pp. 227–328. Academic Press, 1998.
113. M.J. Nieves-Remacha, A.A. Kulkarni, and K.F. Jensen, *Industrial & Engineering Chemistry Research*, Vol. 54, p. 7543, 2015.

114. K.S. Elvira, X.C. Solvas, R.C.R. Wootton, and A.J. deMello, *Nature Chemistry*, Vol. 5, p. 905, October 2013.
115. A. Adamo, R.L. Beingessner, M. Behnam, J. Chen, T.F. Jamison, K.F. Jensen, J.-C.M. Monbaliu, A.S. Myerson, E.M. Revalor, D.R. Snead, T. Stelzer, N. Weeranoppanant, S.Y. Wong, and P. Zhang, *Science*, Vol. 352, p. 61, 2016.
116. S. Rossi, R. Porta, D. Brenna, A. Puglisi, and M. Benaglia, *Angewandte Chemie*, Vol. 129, p. 4354, 2017.
117. Wikipedia contributors, Nitroaldol reaction — Wikipedia, the free encyclopedia, https://en.wikipedia.org/w/index.php?title=Nitroaldol_reaction&oldid=808545871, 2017. [Online; accessed 8-August-2018].
118. J.N. Stuecker, J. Cesarano, III, and J.E. Miller, Regenerable particulate filter, US Patent 7 527 671, assigned to Sandia Corporation (Albuquerque, NM), May 5, 2009.
119. H. Thakkar, S. Eastman, A. Hajari, A.A. Rownaghi, J.C. Knox, and F. Rezaei, *ACS Applied Materials & Interfaces*, Vol. 8, p. 27753, 2016.
120. H. Thakkar, S. Eastman, A. Al-Mamoori, A. Hajari, A.A. Rownaghi, and F. Rezaei, *ACS Applied Materials & Interfaces*, Vol. 9, p. 7489, 2017.
121. H. Thakkar, S. Eastman, Q. Al-Naddaf, A.A. Rownaghi, and F. Rezaei, *ACS Applied Materials & Interfaces*, Vol. 9, p. 35908, 2017.
122. H. Thakkar, S. Lawson, A.A. Rownaghi, and F. Rezaei, *Chemical Engineering Journal*, Vol. 348, p. 109, 2018.
123. Y. Li, S. Chen, X. Cai, J. Hong, X. Wu, Y. Xu, J. Zou, and B.H. Chen, *J. Mater. Chem. A*, Vol. 6, p. 5695, 2018.
124. K.K. Parker, M.R. Badrossamay, and J.A. Goss, Methods and devices for the fabrication of 3D polymeric fibers, US Patent 9 410 267, assigned to President and Fellows of Harvard College (Cambridge, MA), August 9, 2016.
125. T. Xie, J. Wu, Q. Zhao, and Z. Fang, Method for preparing 3D polymer objects with surface microstructures, US Patent Application 20 170 217 079, assigned to Zhejiang University, August 3, 2017.
126. W.-S. Guan, H.-X. Huang, and A.-F. Chen, *Journal of Micromechanics and Microengineering*, Vol. 25, p. 035001, 2015.
127. A.M. Lyons and Q. Xu, Polymer having superhydrophobic surface, US Patent 9 040 145, assigned to Research Foundation of the City University of New York (New York, NY), May 26, 2015.
128. B.G. Compton and J.A. Lewis, *Advanced Materials*, Vol. 26, p. 5930, 2014.
129. N. Grimmelsmann, M. Kreuziger, M. Korger, H. Meissner, and A. Ehrmann, *Rapid Prototyping Journal*, Vol. 24, p. 166, 2018.
130. N. Grimmelsmann, Y. Martens, P. Schäl, H. Meissner, and A. Ehrmann, *Procedia Technology*, Vol. 26, p. 66, 2016. 3rd International Conference on System-Integrated Intelligence: New Challenges for Product and Production Engineering.

131. K. Mondal, *Inventions*, Vol. 3, 2018.
132. D. Brosteaux, F. Axisa, M. Gonzalez, and J. Vanfleteren, *IEEE Electron Device Letters*, Vol. 28, p. 552.554, 2007.
133. M. Sibinski, M. Jakubowska, and M. Sloma, *Sensors*, Vol. 10, p. 7934, 2010.
134. G. Mattana, T. Kinkeldei, D. Leuenberger, C. Ataman, J.J. Ruan, F. Molina-Lopez, A.V. Quintero, G. Nisato, G. Tröster, D. Briand, et al., *IEEE Sens. J.*, Vol. 13, p. 3901, 2013.
135. M. Krehel, M. Wolf, L.F. Boesel, R.M. Rossi, G.-L. Bona, and L.J. Scherer, *Biomed. Opt. Express*, Vol. 5, p. 2537, Aug 2014.
136. C. Cochrane, S.R. Mordon, J.C. Lesage, and V. Koncar, *Materials Science and Engineering: C*, Vol. 33, p. 1170, 2013.
137. R. McLaren, F. Joseph, C. Baguley, and D. Taylor, *Journal of NeuroEngineering and Rehabilitation*, Vol. 13, p. 59, Jun 2016.
138. D. Tosi, S. Poeggel, I. Iordachita, and E. Schena, Fiber optic sensors for biomedical applications in H. Alemohammad, ed., *Opto-Mechanical Fiber Optic Sensors*, chapter 11, pp. 301–333. Elsevier, New York, NY, USA, 2018.
139. D. Morris, S. Coyle, Y. Wu, K.T. Lau, G. Wallace, and D. Diamond, *Sensors and Actuators B: Chemical*, Vol. 139, p. 231, 2009. EUROPT-(R)ODE IX Proceedings of the 9th European Conference on Optical Chemical Sensors and Biosensors.
140. F. Haghdoost, V. Mottaghitalab, and A.K. Haghi, *Sensor Review*, Vol. 35, p. 20, 2015.
141. Q. Huang, L. Liu, D. Wang, J. Liu, Z. Huang, and Z. Zheng, *J. Mater. Chem. A*, Vol. 4, p. 6802, 2016.
142. J. Joseph, S.V. Nair, and D. Menon, *Nano Letters*, Vol. 15, p. 5420, 2015.
143. Y.J. Yun, W.G. Hong, W.-J. Kim, Y. Jun, and B.H. Kim, *Advanced Materials*, Vol. 25, p. 5701, 2013.
144. L. Capineri, *Procedia Engineering*, Vol. 87, p. 724 , 2014. EUROSENSORS 2014, the 28th European Conference on Solid-State Transducers.
145. A. Nilghaz, D.R. Ballerini, and W. Shen, *Biomicrofluidics*, Vol. 7, p. 051501, 2013.
146. S. Choi, S. Kwon, H. Kim, W. Kim, J.H. Kwon, M.S. Lim, H.S. Lee, and K.C. Choi, *Scientific Reports*, Vol. 7, p. 6424, 2017.
147. F. Wienforth, A. Landrock, C. Schindler, J. Siegert, and W. Kirch, *The Journal of Clinical Pharmacology*, Vol. 47, p. 653, 2007.
148. P.K. Sehgal, R. Sripriya, and M. Senthilkumar, Drug delivery dressings in S. Rajendran, ed., *Advanced Textiles for Wound Care*, Woodhead Publishing Series in Textiles, chapter 9, pp. 223–253. Woodhead Publishing, 2009.

149. P. Mostafalu, G. Kiaee, G. Giatsidis, A. Khalilpour, M. Nabavinia, M.R. Dokmeci, S. Sonkusale, D.P. Orgill, A. Tamayol, and A. Khademhosseini, *Advanced Functional Materials*, Vol. 27, p. 1702399, 2017.
150. M. Caldara, C. Colleoni, E. Guido, V. Re, and G. Rosace, *Sensors and Actuators B: Chemical*, Vol. 222, p. 213, 2016.
151. I. Gualandi, M. Marzocchi, A. Achilli, D. Cavedale, A. Bonfiglio, and B. Fraboni, *Scientific Reports*, Vol. 6, p. 33637, September 2016.
152. D. Schmelzeisen, H. Koch, C. Pastore, and T. Gries, 4D textiles: Hybrid textile structures that can change structural form with time by 3D printing in Y. Kyosev, B. Mahltig, and A. Schwarz-Pfeiffer, eds., *Narrow and Smart Textiles*, pp. 189–201. Springer International Publishing, Cham, 2018.

2

Polymers

A description of 3D printing materials, such as filaments, functionalized inks, and powders, can be found in a monograph (1).

2.1 Polymer Matrix Composites

The use of 3D printing for rapid tooling and manufacturing can produce components with complex geometries according to their computer designs (2). Due to the intrinsically limited mechanical properties and functionalities of printed pure polymer parts, there is a critical need to develop printable polymer composites with high performance.

There are many advantages of 3D printing for the fabrication of composites, including high precision, cost-effectiveness, and customized geometry.

The 3D printing techniques of polymer composite materials and the properties and the performance of these composite materials have been reviewed (2). Also, their potential applications in the fields of biomedicine, electronics and aerospace engineering have been detailed.

Thermoplastic polymer materials, such as acrylonitrile-butadiene-styrene (ABS) (3–5), poly(lactic acid) (PLA) (4–6), poly(amide) (PA) (7) and poly(carbonate) (PC) (8), as well as thermosetting polymer materials such as epoxy resins can be processed using a 3D printing technology.

Epoxy resins are reactive materials that require thermal or UV-assisted curing to complete the polymerization process, and they initially exhibit a low viscosity, which rises as the curing proceeds (9–11). Therefore, epoxy resins are suitable for heat or UV-assisted printing process. Based on the various selections of materials, 3D printing of polymers has found possible applications in aerospace industries for creating complex lightweight structures (12), architectural industries for structural models (13), art fields for artifact replication or education (14), and medical fields for printing tissues and organs (15).

Clay-based nanoscale filler materials are commonly used to impart unique and desirable properties to polymer resins (16). Small volume fractions of nanoclay have disproportionately large effects on stiffness, toughness, strength, and gas barrier properties of polymer matrices due to their high surface-to-volume ratio and platelet morphology.

It has been suggested that highly loaded epoxy/clay/fiber mixtures possess desirable rheological properties for their use as feedstock materials for direct-write 3D printing (16).

The effects of a functionalized nanoclay on the rheological properties and printing behavior of an epoxy resin in the absence of fiber reinforcements have been assessed. Also, the effects of clay content and the deposition process on the thermomechanical properties of the resulting 3D printed epoxy/clay nanocomposites were investigated. Flexural strength values range from 80 *MPa* to 100 *MPa* for cast samples and printed samples tested transverse to the printing direction, and up to 143 *MPa* for printed samples tested parallel to the print direction. The strength in each direction is significantly greater than the values for 3D printed thermoplastic composites. This suggests that the epoxy/clay system has high potential for further development as a 3D printing feedstock material (16).

However, most of the 3D printed polymer products are still currently used as conceptual prototypes rather than functional components, since pure polymer products built by 3D printing lack strength and functionality as fully functional and load-bearing parts. Such drawbacks restrict the wide industrial application of 3D printed polymers (2).

The usage of polymer composites for 3D printing can solve the above-mentioned problems by combining the matrix and reinforcements to achieve a system with more useful structural or functional properties not attainable by any of the constituents alone (17).

Incorporation of particle, fiber or nanomaterial reinforcements into polymers permits the fabrication of polymer matrix composites, which are characterized by high mechanical performance and excellent functionality. Conventional fabrication techniques of composites, such as molding, casting and machining, create products with complex geometry through material removal processes (18). While the manufacturing process and performance of composites in these methods are well-controlled and understood, the ability to control the complex internal structure is limited. 3D printing is able to fabricate complex composite structures without the typical waste. The size and geometry of composites can be precisely controlled with the help of computer-aided design. Thus, 3D printing of composites attains an excellent combination of process flexibility and high performance products (2).

Printing techniques and typical polymer materials used therein are shown in Table 2.1.

Table 2.1 Printing techniques and typical polymers (2).

Technique	Typical polymer materials
FDM	Thermoplastics such as PC, ABS, PLA, and nylon
SLA	Photocurable resin (epoxy- or acrylate-based resin)
SLS	PCL and PA powder
3DP	Powder, any materials, binder needed
3D	Liquid or paste PCL, PLA, hydrogel

2.1.1 Biocomposite Filaments

Fully degradable biocomposites for 3D printing applications based on microcrystalline cellulose reinforced PLA have been reported (19).

The biocomposites were produced in filament form by solvent casting and twin screw extrusion to achieve final concentrations of 1%, 3%, and 5% of cellulose. In order to improve the compatibility with the PLA, the cellulose was surface-modified using a titanate coupling agent.

The influence of the cellulose content and modification on the morphological, mechanical, and thermal properties of the biocomposites were studied. Differential scanning calorimetry measurements showed an increase in crystallinity for all biocomposites. The samples with 3 wt% surface-modified cellulose displayed the highest values.

Dynamic mechanical thermal analysis showed that the storage modulus increased for all biocomposites in comparison to neat PLA, with the most significant increase associated with the 3 wt% modified cellulose.

The surface modification of cellulose shifted the tan δ peak of the 1% and 3% biocomposite toward lower temperatures, indicating an increased mobility of the PLA chains. The extruded cellulose reinforced PLA filaments could be successfully 3D printed using a fused deposition modeling technique (19).

2.1.2 *Nanocomposites*

Nanocomposites based on a UV curable polymeric resin and different inorganic fillers were developed for use in UV-assisted 3D printing (20). This technology consists of the additive multilayer deposition of a UV curable resin for the fabrication of 3D macrostructures and microstructures of arbitrary shapes.

A systematic investigation on the effect of the concentration of the filler on the rheological properties of the polymer-based nanocomposites was performed. In particular, the rheological characterization of these nanocomposites allowed the identification of the optimal printability parameters for these systems based on the shear rate of the materials at the extrusion nozzle.

In addition, photocalorimetric measurements were used to assess the effect of the presence of the inorganic fillers on the thermodynamics and kinetics of the photocuring process of the resins. By direct deposition of homogeneous solvent-free nanocomposite dispersions of different fillers in a UV curable polymeric resin, the effect of UV-3D printing direction, fill density, and fill pattern on the mechanical properties of UV-3D printed specimens was investigated by means of uniaxial tensile tests.

Examples of 3D macroarchitectures and microarchitectures, spanning features, and planar transparent structures directly formed upon UV-3D printing of such nanocomposite dispersions could be reproducibly obtained. In summary, the study demonstrated the suitability of these nanocomposite formulations for advanced UV-3D printing applications (20).

2.1.2.1 *Cellulose Nanocrystals*

Cellulose nanocrystals with more than 2000 photoactive groups on each one can act as highly efficient initiators for radical polymerization, as crosslinkers, as well as covalently embedded nanofillers for nanocomposite hydrogels (21).

This can be achieved by a simple method for surface modification of cellulose nanocrystals with a photoactive bis(acyl)phosphane oxide derivative.

Bis(acyl)phosphane oxides efficiently absorb light in the visible range and are therefore interesting photoinitiators widely used in industry.

It was possible to attach bis(acyl)phosphane oxide functions on the surface of cellulose nanocrystals which allowed growing polymer chains from the surface and the preparation of hybrid materials.

Shape-persistent and free-standing 3D structured objects were printed with a monofunctional methacrylate, showing a superior swelling capacity and improved mechanical properties.

2.1.3 Nanowires

2.1.3.1 *Boehmite-Acrylate Composites*

A strategy to obtain a significant enhancement of the mechanical properties for resins via incorporation of boehmite nanowires for surface modification has been reported (22).

Nanowires with a length up to several micrometers and a diameter of around 10 *nm* were obtained by a hydrothermal process (23). The hydrothermal synthesis method is a method of synthesis of single crystals that depends on the solubility of minerals in hot water under high pressure. The crystal growth is performed in an apparatus consisting of an autoclave, in which a nutrient is supplied along with water. A temperature gradient is maintained between the opposite ends of the growth chamber. At the hotter end the nutrient solute dissolves, while at the cooler end it is deposited on a seed crystal, growing the desired crystal (24).

As modifier for the surface modification 2-carboxyethyl acrylate was used, instead of the commonly used coupling agent 3-(trimethoxysilyl) propyl methacrylate. Superior mechanical and rheological properties for 3D printing were obtained.

It was demonstrated that the size and shape of the nanofillers have an impact on the final mechanical properties of the composites. A significant mechanical enhancement could be demonstrated. The superior mechanical enhancement of the composites was attributed to the strong interaction between the boehmite nanowires and 2-carboxyethyl acrylate (22).

2.1.3.2 *Shear-Induced Alignment of Nanowires*

Composite materials achieved via 3D printing offer an opportunity to combine the desired properties of composite materials with the flexibility of additive manufacturing in geometric shape and complexity (25).

The shear-induced alignment of aluminum oxide nanowires during stereolithography printing could be utilized to fabricate a nanowire reinforced polymer composite.

In order to align the fibers, a lateral oscillation mechanism was implemented and combined with wall pattern printing technique to generate a shear flow in both vertical and horizontal directions. A series of specimens were fabricated for testing the tensile strength of the composite material. The results showed that the mechanical properties of the composite were improved by reinforcement of nanofibers through shear-induced alignment. The improvement of tensile strength was approximately 28% by aligning the nanowires at a 5% (1.5% by volume fraction) loading of the aluminum oxide nanowires (25).

2.1.4 *Fiber Reinforced Polymers*

Additive manufacturing is a cost-effective approach for small-scale production. The method provides a higher design flexibility and less material waste than traditional manufacturing techniques (26).

Polymers constitute one of the most popular additive manufacturing materials, yielding lightweight but inherently weak components that cannot hold up against high tension and bending stresses.

A need for improved tensile strength has driven recent interest in additive manufacturing of fiber reinforced polymers.

Additive manufactured fiber reinforced polymers reinforced with short fibers have demonstrated increased mechanical strength, but with limited design and structural flexibility. Additive manufactured fiber reinforced polymers reinforced with continuous fibers provide structural reinforcement within plane. As such, the fibers cannot be extruded along a contoured profile, significantly minimizing the application space for these additive manufactured fiber reinforced polymer devices. This gap was addressed through the development of a new fiber reinforced polymer additive manufacturing process that is capable of continuous fiber deposition along contoured trajectories (26).

The base polymer used in the study was poly(caprolactone) (PCL) (26). A 3D printing system has been detailed that enables the placement of continuous fiber reinforced polymer filaments onto contoured surfaces (e.g., XYZ fiber placement in real time), thereby allowing the user to create shell structures similar to the structures fabricated with traditional manufacturing approaches. One important advantage of 3D printing these structures is the ability to customize the designs on a case-by-case basis (26).

2.1.5 Carbon Fiber Polymer Composites

A class of additively manufactured carbon fiber reinforced composite (AMCFRC) materials has been described (27). The materials have been formed by the use of a latent thermal cured aromatic thermoset resin system, through an adaptation of direct ink writing 3D printing technology.

The following materials were used: Oligomerized bisphenol F diglycidyl ether (BPF) and diethyltoluenediamine (DETDA), Novoset 280 cyanate ester resin, carbon fibers of grade HTS40, and CAB-O-SIL TS-530 fumed silica. The carbon fibers had a fiber tensile modulus of 240 GPa, 1.8% elongation at break and a mean density of 1.77 $g\,cm^{-1}$.

The thermoset carbon fiber composites allow the fiber component of a resin and carbon fiber fluid to be aligned in three dimensions via controlled microextrusion. Subsequently, they can be cured into complex geometries. A high order of fiber alignment within the composite microstructure could be demonstrated. This allows these materials to outperform equivalently filled randomly oriented carbon fiber and polymer composites.

Furthermore, the composite systems exhibit highly orthotropic mechanical and electrical responses as a direct result of the alignment of carbon fiber bundles in the microscale range. The carbon fiber/polymer hybrid materials have locally programmable complex electrical, thermal and mechanical responses (27).

2.1.5.1 Thermosetting Composites

Continuous carbon fiber reinforced thermosetting composites are used in aeronautic and astronautic applications because of their high specific strength, high specific stiffness, good fatigue performance and good corrosion resistance (28,29).

However, the high costs of manufacturing these materials limits their application in the automotive and consumer product industries. Therefore, a more widespread use of these materials will depend on the development of new low-cost manufacturing methods for producing composite structures (30).

The interlaminar deformation and delamination of carbon/epoxy laminated curved beams with variable curvature and thickness were studied experimentally. Both the strength characterization and the evolution of deformation of the laminated curved beams were investigated using a four-point bending test and the digital speckle correlation method (30).

The deformation and fracture behavior of laminated curved beams with various thicknesses and radius-thickness ratios were studied. Some important conclusions are (29):

1. The curved beam strength of laminated curved beam increases with an increase of the thickness and R/t, but the maximum radial stress for laminated curved beams decreases with an increase of the thickness and R/t. The influence of the thickness on the curved beam strength and maximum radial stresses for composite curved beams is larger than that of the radius-thickness ratio.
2. Typical y-strain field of the laminated curved beams was obtained, and the strain measured by the digital speckle correlation method is in accordance with those from the strain

gauges. The crack initiates in the region where the interlaminar strain measured by the digital speckle correlation method is maximum, which can successfully predict the location of crack initiation.

3. Fracture load and failure strain were determined for different specimens, and the different fracture processes of specimens with different thicknesses were also obtained using a charge coupled device camera. The fracture is initiated and propagated in the $(0°, 0°)$ orientation interface, whose critical energy release rate is smallest.

The 3D printed continuous carbon fiber reinforced thermosetting composites were produced and characterized. Some important conclusions are summarized below (30):

1. The 3D printing platform for preparing continuous carbon fiber reinforced thermosetting composites was based on the results of fused deposition modeling (FDM), which dictated the use of a printing head, fiber bundle conveying pipe, epoxy pool, control system, building platform, X-Y motion mechanism, etc. This modified 3D printer can manufacture the composite components over three axes with position control of rotation angles.
2. The composite lamina and composite grid as well as nuts and honeycomb were manufactured using FDM based on the 3D printing platform. The path of the printing head for the composite structures was designed and optimized to produce 3D printing of the continuous carbon fiber reinforced thermosetting composite structures, which was more efficient and affordable than conventional manufacturing techniques.
3. The tensile strength and elastic modulus of the resulting reinforced thermosetting composites was 792.8 MPa and 161.4 GPa. In addition, the flexural strength and elastic modulus were 202.0 MPa and 143.9 GPa, respectively. The mechanical properties of these printed thermosetting composites were better than similar printed reinforced thermoplastic composites and 3D printed short carbon fiber reinforced composites.

2.1.6 *FDM Printing*

The most commonly used printers for fabricating polymer composites are FDM printers (2). Here, thermoplastics such as PC, ABS and PLA are commonly used due to their low melting temperature. These printers work by controlled extrusion of thermoplastic filaments. In FDM, filaments melt into a semi-liquid state at the nozzle and are extruded layer by layer onto the build platform. Here the layers are fused together and then solidify into the final parts. The quality of printed parts can be controlled by altering the parameters of printing, such as layer thickness, printing orientation, raster width, raster angle and air gap. The effects of processing parameters have been discussed (31).

The strength of a FDM processed component primarily depends on five important control factors such as layer thickness, part build orientation, raster angle, raster width, and raster to raster gap, i.e., air gap. These parameters are defined as (32):

1. Orientation: Part build orientation or orientation refers to the inclination of part in a build platform with respect to X, Y, Z axis. X- and Y-axis are considered parallel to build platform and Z-axis is along the direction of part build.
2. Layer thickness: It is a thickness of layer deposited by nozzle and depends upon the type of nozzle used.
3. Raster angle: It is a direction of raster relative to the x-axis of the build table.
4. Part raster width (raster width): Width of raster pattern used to fill interior regions of part curves.
5. Raster to raster gap (air gap): It is the gap between two adjacent rasters on the same layer.

A common drawback of FDM printing is that the composite materials have to be in a filament form to allow the extrusion process. However, it is difficult to homogeneously disperse reinforcements and to remove the void that is formed during the manufacturing of composite filaments. Another disadvantage of FDM printers is that the usable material is limited to thermoplastic polymers with a suitable melt viscosity. The molten viscosity should be high enough to provide structural support and low enough to enable the extrusion process. Also, the complete removal of the support structure used during printing may be difficult.

However, FDM printers also have advantages, including low cost, high speed, and simplicity. Another advantage of FDM printing is the potential to simultaneously allow the deposition of diverse materials. Multiple extrusion nozzles with loadings of different materials can be set up in FDM printers, so the printed parts can be multifunctional with a designed composition (2).

2.1.6.1 *Modeling*

The FDM printing process of continuous fiber reinforced composites exhibits a typical fluid-structure interaction problem, where fibers deform and deposit in the melted plastic, which is extruded as a fluid (33).

Preciously, particle models have been developed for both fluid-solid flow (34) and fluid-structure interaction (35), which allows the free surface flow of melted plastic, fiber-resin interaction, and fiber collision/failure. In a more recent study an attempt was made to model the 3D printing process of composites (33).

A numerical approach has been presented for modeling the 3D printing process of fiber reinforced polymer composites by FDM (33). This approach was based on the coupling between two particle methods, i.e., smoothed particle hydrodynamics and the discrete element method.

In smoothed particle hydrodynamics, the resin in fluid phase is spatially represented by a set of discrete particles moving in accordance with the Navier-Stokes equations. The movement of each particle is determined by the overall interactions from nearby particles within a support domain. Each particle, which carries its own mass and density, is affected by the surrounding particles within the support domain. The density of each particle can be approximated by using a method called continuity density, in which the approximation of density is processed through the continuity equations.

$$\frac{D\rho_i}{Dt} = \sum_{j=1}^{N} m_j v_{i,j} \partial \frac{\partial W_{i,j}}{\partial x_j} \tag{2.1}$$

Here, i is the observed particle, j are the surrounding particles, $v_{i,j}$ are the relative velocities, m is the mass, $W_{i,j}$ is the kernel function and its gradient determines the contribution of the relative velocities.

In the same way as the particle approximation of the density in Eq. 2.1, the moment equation in smoothed particle hydrodynamics can be formed (33).

The drag force is the only force hydrodynamically acting on the particles (solid phase, fibers), using the discrete element method. The motion of each particle is governed by various forces, e.g., drag force, bond force and direct contact force. Then its reaction is returned to the smoothed particle hydrodynamics particles (fluid phase: resin) accordingly.

The so coupled model has distinctive advantages in dealing with the free surface flow, large deformation of fibers, and/or fiber-fiber interaction that are involved in the FDM process.

A numerical feasibility study was carried out to demonstrate its capability for both short and continuous fiber reinforced polymer composites, with promising results achieved for the rheological flow and fiber orientation and deformation (33).

Metal/Polymer Composite Filaments. Metal/polymer composite filaments for FDM processes have been developed in order to observe the thermomechanical properties of these filaments (36). An ABS thermoplastic was mixed with copper and iron particles. The percent loading of the metal powder was varied to confirm the effects of metal particles on the thermomechanical properties of the filament, such as tensile strength and thermal conductivity.

The printing parameters, such as temperature and fill density, were also varied to assess the effects of the parameters on the tensile strength of the final product which was made using the FDM process.

It could be confirmed that the tensile strength of the composites is decreased by increasing the loading of metal particles. In addition, the thermal conductivity of the metal/polymer composite filament was found to be improved by increasing the metal content.

Therefore, the metal/polymer filament could be used to print metal and large-scale 3D structures without any distortion by the thermal expansion of the thermoplastics. The material could also be used in 3D printed circuits and electromagnetic structures for shielding and other applications (36).

2.1.7 *Powder Bed and Inkjet Head 3D Printing*

The powder-liquid 3D printing technology was developed at the Massachusetts Institute of Technology in 1993 as a rapid prototyping technology (37). This technique is based on powder processing. The powders are first spread on the build platform and then selectively joined into a patterned layer by depositing a liquid binder through an inkjet printhead. The printhead can move in the X-Y direction.

After a desired two-dimensional pattern is formed, the platform lowers and the next layer of powder is spread. The process is repeated and finally unbounded powder must be removed to get the final products. The internal structure can be controlled by altering the amount of the deposited binder. The factors determining the quality of final products are powder size, binder viscosity, interaction between binder and powder, and the binder deposition speed (38).

The major advantages of this technology are the flexibility of material selections and the possibility of room temperature processing. Basically, any polymer materials in powder state can be printed by this technology. The removal of the support structure is comparatively easy. However, the binder used may incorporate other contaminations and the printing resolution is also very limited for this technology (2).

2.1.8 *Stereolithography*

The stereolithography (SLA) technique uses photopolymers that can be cured with a UV laser. Here, the photocurable resin will polymerize into a 2D patterned layer. After each layer is cured, the platform lowers and another layer of uncured resin is ready to be patterned (39). Typical polymer materials used in SLA are acrylic resins and epoxy resins.

The curing reactions that occur during polymerization must be fully understood to control the quality of the final printed parts. The intensity of laser power, scan speed and duration of exposure affect the curing time and printing resolution (40).

In micro-stereolithography, the photopolymer solidification is affected by fabrication conditions, such as the optical properties, i.e., laser power, laser scanning speed, laser scanning pitch focusing condition, etc., and the material properties of the photopolymer. Thus, the photopolymer solidification phenomena must be considered when generating a laser scanning path. A scanning path generation algorithm that uses 3D computer-aided design (CAD) data and considers the photopolymer solidification phenomena has been proposed to improve the dimensional accuracy in micro-stereolithography (40). Photoinitiators and UV absorbers can be added to the resin to control the depth of polymerization (41).

The main advantage of SLA printing technology is the ability to print parts with high resolution. Additionally, because SLA is a nozzle-free technique, the problem of nozzle clogging can be avoided. Despite these advantages, the high cost of this system is a main concern for industrial application. Possible cytotoxicity of residual photoinitiator and uncured resin is another concern (2).

2.1.9 *Selective Laser Sintering*

The selective laser sintering (SLS) technique is similar to the three-dimensional printing 3DP technique. Both are based on powder processing. Instead of using a liquid binder, in SLS, a laser beam with a controlled path scans the powders to sinter them by heating.

Using high power lasers, neighboring powders are fused together through molecular diffusion and then processing of the next layer starts. Unbounded powder should be removed to get the final products (42).

The feature resolution is determined by powder particle size, laser power, scan spacing and scan speed (43).

2.2 Sequential Interpenetrating Polymer Network

A sequential interpenetrating polymer network has been obtained by the formulation of a photocurable acrylic resin with a thermocurable epoxy resin (44). This composition was proposed as a matrix for the fabrication of carbon fiber reinforced composite structures by means of a 3D printing technology.

This approach combines the advantages of the easy free-form fabrication typical of the 3D printing technology with the purposely customized features of the interpenetrating polymer network material.

The following materials were used in the study: Bisphenol A diglycidyl ether (DGEBA), 1,1-dimethyl,3-(3′,4′-dichlorophenyl)urea (Diuron™), dicyandiamide (DICY), and fumed silica (200 $m^2 g^{-1}$, primary particle size distribution 8–20 *nm*, ox200) from Sigma-Aldrich were used. Bisphenol A ethoxylate diacrylate (SR349) was provided by Arkema, whilst the photoinitiator 2,4,6-trimethylbenzoylphenyl phosphinate (Irgacure TPO-L) was obtained from BASF. Carbon fibers were provided by Zoltek (Panex™ 30 milled carbon fibers, 99% carbon, density 1.8 $g\,ml^{-1}$, fiber diameter 7.2 μm, fiber length 100–150 μm).

Formulations were obtained by liquid blending the photocurable acrylic resin and the thermocurable epoxy resin with the suspended hardener (DICY) and the accelerator (Diuron).

Photocalorimetric and dynamic mechanical characterization were performed in order to investigate the photo- and thermo-crosslinking reactions and their effect on the development of the interpenetrating polymer network system. The interpenetrating polymer network resin was finally loaded with carbon fibers and successfully ultraviolet-assisted 3D printed. This demonstrated the possibility of fabricating carbon fiber reinforced composite materials in 3D with excellent mechanical properties.

The study clearly showed the great potential of this additive manufacturing technology for advanced industrial applications (44).

2.3 3D Printable Diamond Polymer Composite

The development of a thermally conducting composite material that can be rapidly 3D printed into prototype objects has been presented (45).

The composite structures contain 10, 20, 25 and 30% w/v of 2–4 micron-sized synthetic diamond microparticles which are added to the acrylate polymer. These compounds were produced using a low-cost stereolithographic 3D printer.

The so prepared materials were characterized with respect to heat transfer rate, thermal expansion coefficient and contact angle.

These composites displayed minor enhancements in the heat transfer rate with incrementing diamond content up to 25% w/v.

However, a significant improvement was observed for a 30% w/v polymer-diamond composite, based on an interconnected diamond aggregate network, as confirmed by high resolution scanning electron microscopy.

The material developed herein was used in the fabrication of prototype 3D printed heat sinks and cooling coils for thermal management applications in electronic and fluidic devices. Infrared thermal imaging performed on 3D printed objects verified the superior performance of the composite in comparison to the inherent polymer (45).

2.4 Adhesives for 3D Printing

A weak bond between the extruded print material, or the support material, if a raft is used, and the surface of the print pad can cause the extruded print material to separate from the print pad prematurely (46). Such separation can lead to a failed print process and/or to imperfections in the printed 3D object. In contrast, a strong bond can prevent the extruded print material from separating from the print pad prematurely but may also make it difficult to remove the printed 3D object and any support material from the print pad following completion of the print process. Furthermore, this difficulty can require a user to employ a hand tool such as a scraper to separate the printed 3D object from the print pad, which can lead to damage to the printed 3D object and/or the surface of the print pad.

An adhesive for use with a 3D printer has been described (46). It consists of a first polymeric component, a poly(vinyl alcohol) and a second polymeric component, a water-soluble polymer such as a poly(2-ethyl-2-oxazoline). In addition, the adhesive composition further contains a solvent, a surfactant, and/or a preservative.

The solvent can be water, including distilled water or deionized water. In some embodiments, a solvent consists of one or more of acetone, ethanol, methanol, ethylene glycol, propylene glycol, triethylene glycol, glycerin, acetamide, dimethyl acetamide, dimethyl sulfoxide, methyl ethyl ketone, methylene chloride, and combinations or mixtures thereof.

An alkoxylated diol can be used as surfactant. Also, a cationic, anionic, or zwitterionic surfactant can be used, e.g., a quaternary ammonium compound or quaternary ammonium salt such as cetyltrimethylammonium bromide. Other surfactants suitable for use include SURFYNOL® 104 (tetramethyl decynediol), SURFYNOL 440 (ethoxylated 2,4,7,9-tetramethyl 5-decyn-4,7-diol), SURFYNOL 2502 (ethoxylated propoxylated 2,4,7,9-tetramethyl-5-decyne-4,7-diol), and DYNOL® 604, available from Air Products (47).

As a preservative an inorganic composition can be used, including metals and/or metal salts. For example, a preservative comprises metallic copper, zinc, or silver or a salt of copper, zinc, or silver (46).

2.5 Voronoi-Based Composite Structures

A novel design, additive manufacturing and modeling approach for three-dimensional Voronoi-based composite structures that closely mimic nacre's multilayer composite structure has been shown (5). The hierarchical structure of natural nacre is mimicked to produce multilayer composite laminates assembled from three-dimensional polygonal tablets bonded with organic adhesives. In addition, various complex geometries of the nacreous shells observed in nature, such as the dome-shaped structure, could be developed into three-dimensional designs. A mapping algorithm has been developed to generate complex structures of nacre-like composites that are readily fabricated by unique dual-material 3D printing technology.

A numerical model of the nacreous composite has been proposed, which includes tablet cohesive bonds and interlaminate adhesive layers to mimic the soft organic polymer matrix. The nacreous model was validated against a natural nacre specimen under uniaxial loading. To exemplify a potential application, a scaled model of a nacre-mimetic composite made of aluminum tablets and a vinylester adhesive was constructed and assessed against blast-induced impulsive loading (5).

2.6 Graphene Oxide Reinforced Complex Architectures

The dependency of polymer-based nanocomposites on distribution, concentration, geometry and property of nanofillers in polymer matrix were elucidated (48).

Increasing the concentration of carbon-based nanomaterials, such as carbon nanotubes (CNTs), in polymer matrix often results in stronger, but more brittle materials.

Three strategies, i.e., *in-situ* intercalative polymerization, solution intercalation, and melt intercalation, have been used to prepare graphene/polymer or a graphene oxide/polymer:

***In-situ* intercalative polymerization:**
In this method, graphene or graphene oxide is first swollen in the liquid monomer, then a suitable initiator is diffused and polymerization reaction proceeds by heat or radiation.

Solution intercalation:
Three steps are involved in this method: Dispersion of graphene or graphene oxide in a suitable solvent, addition of polymer and removal of the solvent. Graphene or graphene oxide can be dispersed in various solvents, which can be water, acetone, chloroform, tetrahydrofuran, dimethyl formamide or toluene, by ultrasonication. Graphene or graphene oxide sheets are delaminated by breaking the weak forces that stack these layers together, and the polymer then is adsorbed onto these sheets. The separated graphene sheets sandwich the polymer to form a nanocomposite by removing solvent, which is crucial for the properties of nanocomposites. Increased entropy by desorption of solvent molecules is the driving force of polymer intercalation, which compensates for the decreasing entropy of intercalating polymer chains. This means that a large quantity of solvent molecules are needed for desorbing from graphene or graphene oxide sheets to accommodate incoming polymer chains. The

main advantage of this method is introducing low or even no polarity during the synthesis process.

Melt intercalation:

Graphene or graphene oxide is mechanically mixed with thermoplastic polymer at elevated temperatures by extrusion or injection molding. Nanocomposites are formed by intercalating or exfoliating polymer chains. This method is solvent free.

In the experiments, a single-layer graphene oxide from Cheap Tubes Inc. (49) was used, and as polymer resin PIC 100 from EnvisionTEC (50) was used.

A three-dimensional printed graphene oxide complex structure was fabricated by stereolithography with a good combination of strength and ductility. With a content of only 0.2% graphene oxide, the tensile strength is increased by 62.2% and the elongation could be increased by 12.8%. Transmission electron microscope measurements showed that the graphene oxides were randomly aligned in the cross section of the polymer.

The strengthening mechanism of the 3D printed structure was investigated in terms of tensile strength and the Young's modulus. It was found that an increase in ductility of the 3D printed nanocomposites is related to an increase in the crystallinity of graphene oxide reinforced polymer. A compression test of the 3D graphene oxides structure revealed the metal-like failure model of graphene oxide nanocomposites (48).

2.7 Multiwalled Carbon Nanotube Composites

Selective laser sintering (SLS) offers a strategy to create 3D complex components with desirable mechanical, electrical and thermal properties using composite powders as feeding materials. A fabrication approach has been proposed to prepare CNT composite powders and to use these materials for an SLS process (51).

In comparison to a hot-compression process, the SLS process offers an effective method to fabricate the CNT/polymer composites with electrically conductive segregated structures.

At a small loading range of CNTs smaller than 1%, the laser sintered composites exhibit a significant improvement in electrical conductivity up to anti-static and conductive range, qualifying the applications in automobile and aerospace. However, the enhancement in the thermal conductivity of such laser sintered composites is not comparable with that of hot-compressed analogues.

A powder-based 3D printing method has been presented for the preparation of electrically conductive components with complex geometries by using a laser to selectively sinter CNT reinforced composite powders (51). A PA12 thermoplastic (PA2200 from EOS GmbH, Krailling, Germany) (52) was used as the polymeric matrix of the composite. The PA12 had a melting point of 184°C and the mean density of its powders was 0.45 $g\,cm^{-3}$. In addition, thermoplastic poly(urethane) (TPU) (Desmosint X92, BASF Germany) (53) was used.

The PA12 and PU powders were dispersed into the deionized water to form an aqueous suspension, which was then heated up to 90°C and 70°C for 30 *min*, respectively. Under the elevated temperatures, the surfaces of the polymeric powders are softened and activated. The multiwalled carbon nanotube suspension was dispersed in deionized water with sodium cholate hydrate and sonicated for 30 *min* to disaggregate the nanotubes and to form a homogeneous suspension. The addition of sodium cholate hydrate was kept saturated with respect to the quantity of the multiwalled carbon nanotubes. Then, the two solutions were combined. Finally, the multiwalled carbon nanotube coated polymeric powders were filtrated and dried. A 3D printing process was conducted on PA12, PU and their CNT reinforced composite materials using an SLS machine. The machine contained a carbon dioxide laser (51).

The relationships of process, structure, and properties, revealed that the laser sintering-induced segregated microstructures can be favorable for electron conduction, but the inevitable pores adversely affect the thermal conductivity of the laser sintered composites. The facile method via laser sintering composite powders could be implemented for the fabrications of different thermoplastic components with segregated conductive microstructures. These robust conductive composite parts (CNT/PA12) and the highly porous hybrid components (CNT/PU) are both potentially applicable in applications for aerospace, automobile, electronic packaging and actuation (51).

2.8 Multifunctional Polymer Nanocomposites

The latest progress in 3D printing of multifunctional polymer nanocomposites and microfiber reinforced composites has been reviewed (54–57).

There are various types of nanomaterials, such as carbon nanotubes, carbon nanofibers, graphene, metallic nanowires, metallic and ceramic nanoparticles, clay nanoplatelets, and quantum dots. The incorporation of these nanomaterials into certain polymers enables the fabrication of polymer nanocomposites. These compositions exhibit unique properties or functionality (57), e.g., electrical and thermal conductivities (58), electromechanical sensitivity (59), magnetism (60), and mechanical strength (61).

The ability to print 3D structures in a layer-by-layer manner enables manufacturing of highly customized complex features and allows efficient control over the properties of the fabricated structures.

The incorporation of various types of nanomaterials into a wide range of polymer matrices, both thermoplastics and thermosetting resins to manufacture structures, has been greatly reported in the literature. Despite the many advantages offered by the nanomaterials, the properties are still below the theoretical predictions due to the many challenges that arise during the nanocomposite processing stages.

The creation of nanocomposites that exhibit the desired properties is complex and requires a tight control over the nanomaterials' dispersion, distribution, orientation, and connection with each other and with the matrix.

Various mixing strategies involving a surface treatment of the nanomaterials, e.g., chemical functionalization of nanomaterials to create stronger connections with the surrounding matrix (62, 63) and their mixing using a sonication bath or a three-roll mill mixer for thermosets (64) or an extruder for thermoplastics (65), can be used to properly disperse the nanomaterials inside the polymer. Various mixing strategies and their effects on the overall properties of the nanocomposites have been reviewed (66). Here, recent studies in polymer nanocomposites based on layered silicate have been presented. The background, morphology, preparation methods, and the properties of these materials were discussed.

The 3D printing techniques that can be used for nanocomposites/composites, and their compatible materials are summarized in Table 2.2.

Table 2.2 3D printing techniques for nanocomposites (57).

Technique	Material
Stereolithography (SLA)	UV curable polymers
Optical projection SLA	UV curable polymers
Two-photon polymerization (TPP)	UV curable polymers
Selective laser sintering (SLS)	Thermoplastics
Fused deposition modeling (FDM)	Thermoplastics
Binder jetting	Solutions/Suspensions
PolyJet	UV curable polymers
UV-3D printing	UV curable polymers
SC-3D printing	Thermoplastic solutions

There has been increasing use of 3D printing of fiber reinforced thermoplastic composites in various domains. For example, high-performance thermoplastic composites such as carbon fiber reinforced poly(ether ether ketone) (PEEK) or poly(phenylene sulfide) have shown huge potential to replace many aerospace metallic parts for energy efficiency and fuel economy. These materials are mechanically strong, rigid, chemically inert, non-flammable, light, and more importantly they retain their properties under harsh environmental conditions, especially at high temperatures of up to 400°C (67,68).

The aerospace industry has already started to replace metallic components of engines with the high-performance composites, and continues to find new applications (69).

As an example, NASA has reported the utilization of poly(ether imide) polymers reinforced with chopped carbon fibers for FDM printing of acoustic materials for engine noise reduction in a single-step manufacturing process (70). This omits additional work such as bonding and drilling associated with the current manufacturing approach. The manufactured honeycomb structure featuring complex geometries performed as well as conventional honeycomb/facesheet liners in wind tunnel tests.

3D printing of composites further enables the improvement of properties with innovative designs to help reduce noise over a specified frequency range in a lighter, compact design (57).

2.9 Additive Manufacturing

Comprehensive reviews of the main 3D printing methods, materials and their development in trending applications has been presented (71,72).

In particular, the revolutionary applications of additive manufacturing in biomedical, aerospace, buildings and protective structures are discussed. Also, the current state of materials development, including metal alloys, polymer composites, ceramics, and concrete, has been detailed (72).

Additive manufacturing alias 3D printing translates CAD virtual 3D models into physical objects (73). By digital slicing of CAD, 3D scan, or tomography data, additive manufacturing builds objects layer by layer without the need for molds or machining.

Additive manufacturing enables the decentralized fabrication of customized objects on demand by exploiting digital information storage and retrieval via the Internet. The ongoing transition from rapid prototyping to rapid manufacturing prompts new challenges for mechanical engineers and materials scientists alike.

Because polymers are by far the most utilized class of materials for additive manufacturing, polymer processing and the development of polymers and advanced polymer systems has been reviewed, specifically for additive manufacturing (73). Additive manufacturing techniques covered include vat photopolymerization (stereolithography), powder bed fusion (SLS), material and binder jetting (inkjet and aerosol 3D printing), sheet lamination (laminated object manufacturing), extrusion (FDM, 3D dispensing, 3D fiber deposition, and 3D plotting), and 3D bioprinting.

The range of polymers used in additive manufacturing encompasses thermoplastics, thermosets, elastomers, hydrogels, functional polymers, polymer blends, composites, and biological systems.

The aspects of polymer design, additives, and processing parameters as they relate to enhancing build speed and improving accuracy, functionality, surface finish, stability, mechanical properties, and porosity have been addressed. Selected applications demonstrate how polymer-based additive manufacturing is being exploited in lightweight engineering, architecture, food processing, optics, energy technology, dentistry, drug delivery, and personalized medicine.

Unparalleled by metals and ceramics, the technique of polymer-based additive manufacturing plays a key role in the emerging additive manufacturing of advanced multifunctional and multimaterial systems, including living biological systems, as well as lifelike synthetic systems (73).

The following methods and the polymers used have been discussed (73):

1. Material extrusion is an additive manufacturing process in which material is selectively dispensed through a nozzle.
2. Fused deposition modeling (FDM), fused filament fabrication (FFF), 3D dispensing, and 3D bioplotting fall into this category.
3. Material jetting is an additive manufacturing process in which droplets of build material (such as photopolymer or thermoplastic materials) are selectively deposited. Systems based on inkjet printing fall into this category.
4. Binder jetting is an additive manufacturing process in which a liquid bonding agent is selectively deposited to fuse powder materials. Sheet lamination is an additive manufacturing process in which sheets of material are bonded together to form an object.
5. Vat photopolymerization is an additive manufacturing process in which liquid photopolymer in a vat is selectively cured by light-activated polymerization. Many of the lithography-based additive manufacturing approaches (e.g., multi-photon polymerization (2PP), digital light processing, and stereolithography (SLA)) can be grouped into this category.
6. Powder bed fusion is an additive manufacturing process in which thermal energy (provided, e.g., by a laser or an electron beam) selectively fuses regions of a powder bed. Both selective laser sintering (from 3D systems) and laser sintering (from EOS), and electron beam machining (EBM) fall into this category. These processes are used for metals as well as polymers.
7. Directed energy deposition is an additive manufacturing process in which focused thermal energy (e.g., laser or

plasma arc) is used to fuse materials by melting as they are being deposited. This process is currently only used for metals.

2.9.1 *Thermosetting Polymers*

Traditional additive manufacturing or FDM relies on thermoplastics. Methods and compositions for additive manufacturing have been described that include reactive or thermosetting polymers such as urethanes and epoxies (74).

Using thermoset materials can minimize the distortion during a build, increase layer-to-layer adhesion, and/or minimize anisotropy in the part that results from poor interlayer adhesion.

Exemplary commercial epoxy precursors include aromatic epoxies, e.g., epoxy bisphenol A or aromatic/aliphatic epoxies (74). For the latter materials, the aromaticity enhances the microstructure development under magnetic fields, while the aliphatic segments lower the viscosity of the epoxy to enhance the processability. The epoxy can be blended with an aromatic amine curing agent to obtain a partially reacted prepolymer at moderate temperatures (near room temperature) and a second less reactive phenolic curing agent for high temperature curing.

The polymers are melted, partially crosslinked prior to the deposition, deposited to form a component object, solidified, and then fully crosslinked. The polymers form networks of chemical bonds that span the deposited layers.

A directional electromagnetic field can be applied to aromatic polymers after the deposition to align the polymers for improved bonding between the deposited layers (74).

2.9.2 *UV Curable Materials*

A resin, which had been originally intended for use as a UV curable adhesive, consisted of a urethane dimethacrylate with a small fraction of acrylic acid, benzophenone as photoinitiator, and methyl ethyl hydroquinone/triallyl phosphate to inhibit a premature polymerization (75).

As photo-based additive manufacturing methods matured, scientists sought to provide resins that required a smaller dose of energy

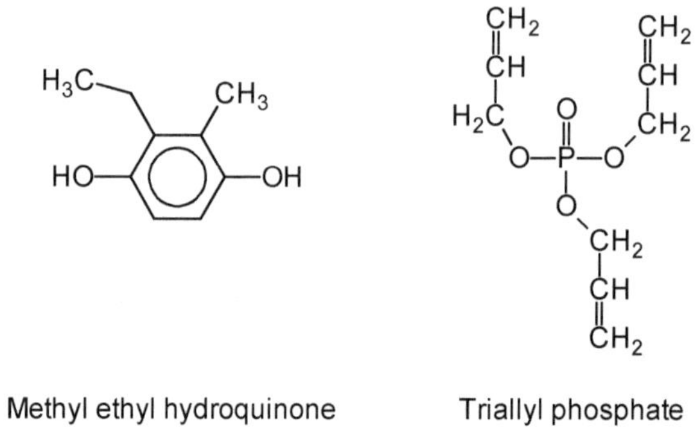

Methyl ethyl hydroquinone   Triallyl phosphate

Figure 2.1 Inhibitors.

to reach gelation, which relate to faster writing speeds and thus rapid prototyping (76,77).

Radical photoinitiators are used for radical generation, initiation, and propagation (73). These are comparatively rapid processes and are thus commonly used for rapid prototyping. The first step in crosslinking of a photoresin is the absorption of light, and the photoinitiator converts the photolytic energy into reactive species to induce the polymerization process. Radical photoinitiators may be classified as either Norrish Type I or Type II (78). Type I initiators are single molecules that cleave into radical fragments when exposed to light of an appropriate wavelength. Benzil ketals are common Type I initiators with fairly low energy $n \rightarrow p^*$ transitions, equivalent to roughly 350–360 *nm*.

Acyl phosphine oxides (such as TPO and BAPO, cf. Figure 2.2) are another class of Type I radical initiators and may be preferred for additive manufacturing devices based on higher wavelength lamps (79). The phosphorus atom adjacent to the carbonyl group lowers the energy level of the p* state, thus shifting the maximum of the $n \rightarrow p^*$ transition toward 400 *nm*. Most essentially, these and the new generation of long wavelength germanium initiators (Ivocerin) have excellent photobleaching behavior, as the heteroatom is separated from the carbonyl group, thus allow curing of highly filled systems (80).

Type II photoinitiation systems are two-component systems consisting of a light-absorbing molecule or a sensitizer, along with a

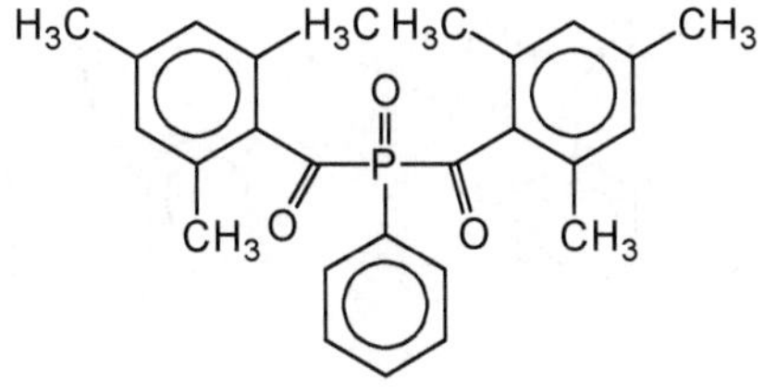

(Phenylphosphoryl)bis(mesitylmethanone)

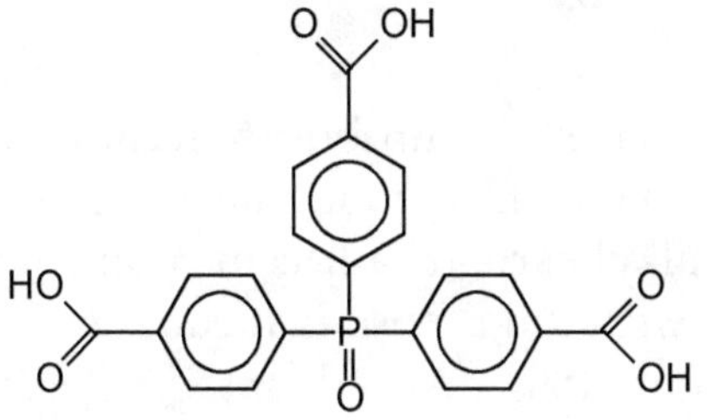

Tris(4-carboxyphenyl)phosphine oxide

Figure 2.2 Acyl phosphine oxides.

co-initiator, or synergist. Upon irradiation, the synergist donates a hydrogen atom to the excited sensitizer and in the process provides the initiating radical. Tertiary amines with at least one alkyl substituent are the most commonly used Type II co-initiators (81,82).

They react via electron transfer from the amine to the excited ketone. The subsequent step of proton transfer from the amine radical cation to the initiator radical anion is the speed limiting factor, and back electron transfer has to be considered as a competitive reaction. Although Type II sensitizers (such as benzophenone and 1-isopropylthioxanthone, cf. Figure 2.3) are commonly cited in SLA patents (83), amine synergists are not recommended. One reason is that many modern day SLA resins are based on mixtures of radical- and cationic-based systems, and amines can inhibit the latter. In fact, benzyl-*N*,*N*-dimethylamine (a very effective hydrogen donor) is listed as a cationic stabilizer in a few patents (84), and used at a concentration of less than 0.01%, which is too low to be of use as a synergist.

While Type II systems are more effective with reactive synergists,

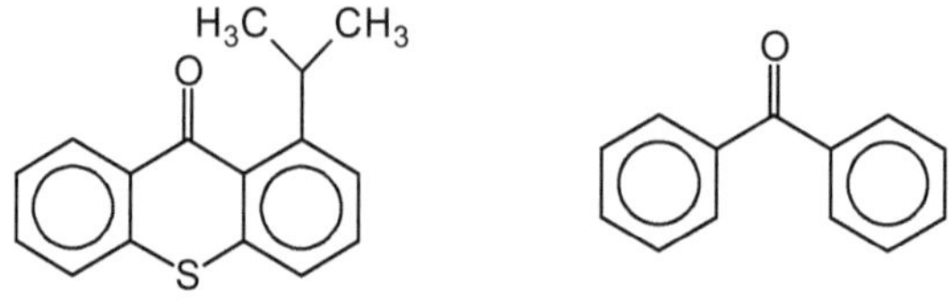

1-Isopropylthioxanthone Benzophenone

Figure 2.3 Type II sensitizers.

Type II sensitizers may still contribute to radical polymerization via direct hydrogen donation from monomer or solvent (85). Hydrogen abstraction from alkyl carbons adjacent to oxygen can occur with many monomers, including those that contain propylene glycol or ethylene glycol units (86). An advantage of using oligoether or oligoester acrylate monomers with abstractable hydrogens is that they tend to cure better in air than analogous alkyl monomers, and yet they do not inhibit cationic polymerization such as amines (73).

2.9.3 *(Meth)acrylate Monomers*

Because the exact composition of most commercial resins used in photo-based additive manufacturing is proprietary, examples from patents are used to provide insight into utilized monomers and potential concentrations thereof. As a good early example of monomers intended for SLA, Murphy *et al.* filed a patent in 1988 describing a resin consisting of a combination of a high viscosity oligomeric diacrylate or dimethacrylate dissolved in a liquid acrylate or methacrylate and an *N*-vinyl monomer (preferably *N*-vinyl pyrrolidone (NVP) as reactive diluent, cf. Figure 2.4) (87).

They state that a system consisting of both an acrylate and a methacrylate is preferable because methacrylates cure too slowly on their own and because the pure acrylate system leads to distortions in the printed object. In their examples, they describe a resinous diacrylate (either urethane- or epoxy-based) dissolved in trimethylol propane trimethacrylate (TTMA) or hexanediol dimethacrylate. NVP is rapid curing and provides green strength, which refers to the combined mechanical properties required to maintain

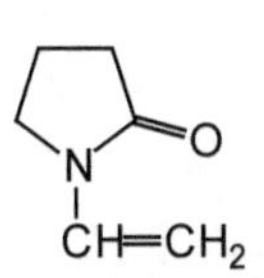

O
N
$CH{=}CH_2$

N-Vinylpyrrolidone

Figure 2.4 Reactive diluent.

fidelity during the development process. The ideal ratio of resinous acrylate:liquid methacrylate:NVP was found between 7:6:6 and 14:3:3. Darocur 1173 (D1173) and Irgacure 184 (I184) were listed as photoinitiators.

Resins based on urethane acrylates and DGEBA are commonly cited in SLA patents due to the mechanical strength that these functional groups help provide (88, 89).

Urethane acrylates are synthesized from the reaction of hydroxy acrylates (such as HEA or HEMA) with isocyanates, where the latter is oftentimes an oligomeric poly(urethane) formed in a prior step from a polyol reacted with excess small molecule diisocyanate (90).

An array of urethane acrylate monomers is available commercially, defined as either aliphatic or aromatic and with degree of functionality from one to six. Structural monomers, on their own, are often too viscous to be processed directly by additive manufacturing and require thinning with a lower viscosity reactive diluent (91).

To facilitate rapid crosslinking of the resin, multifunctional reactive diluents such as dipropylene glycol diacrylate (DPGDA) or pentaerythritol tetraacrylate (PETA) are often used. Tris[2-(acryloyl)ethyl] isocyanurate (TAEI), cf. Figure 2.5, is a reactive liquid acrylate monomer with a heterocycle core that should also help improve product mechanical properties (92).

Diacrylates with cycloaliphatic cores (such as DCPDA) are claimed in a few SLA patents as they tend to undergo less shrinkage than other acrylate monomers and help contribute to a higher final modulus (93).

Although they have their advantages, acrylates, as with all other vinyl monomers, undergo shrinkage during polymerization. The amount of shrinkage is dependent on molecular structure, with cycloaliphatic and aromatic acrylates shrinking less than common

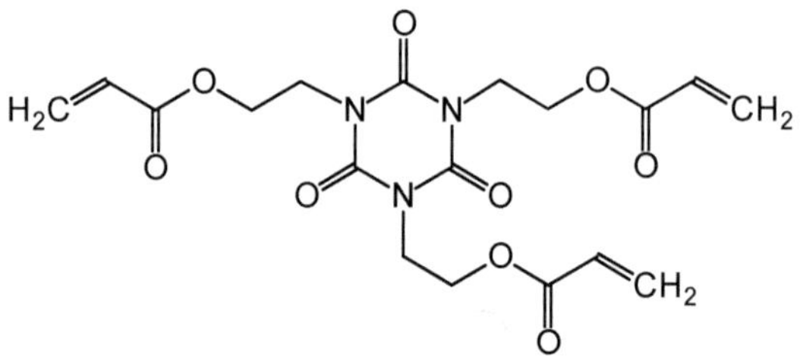

Tris[2-(acryloyl)ethyl] isocyanurate

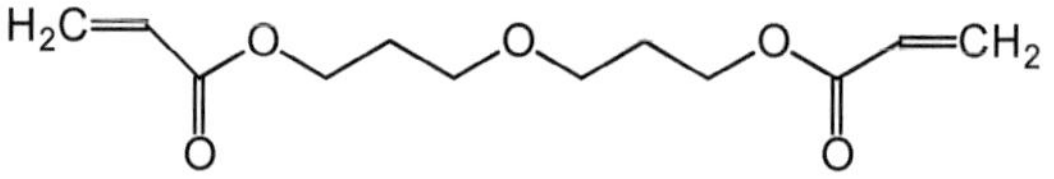

Dipropylene glycol diacrylate

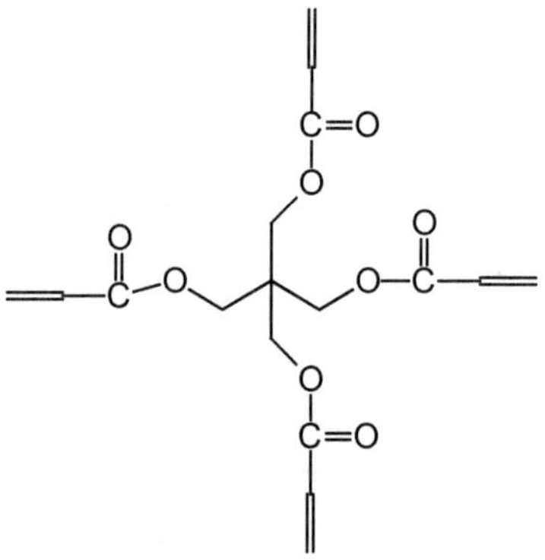

Pentaerythritol tetraacrylate

Figure 2.5 Multifunctional reactive diluents.

diluents (i.e., bisphenol-based dimethacrylate bis-GMA shrinks 5%, while diluent triethylene glycol dimethacrylate shrinks 12%) (94).

Preorganization of monomers (e.g., by hydrogen bridges) can help to reduce shrinkage stress. Another strategy is to change the polymerization mechanism from a chain-growth polymerization toward a radical step-growth mechanism (elaborated further in subsequent sections). Polymerization shrinkage and associated stress cause particular problems in layer-by-layer fabrication where inhomogeneous stress results in curling and other deformation problems (95). Hull *et al.* describe 3D printing techniques including the use of dashed and curved lines for vertical structures, which develop less strain versus straight continuous structures. One of the more common chemical methods of reducing shrinkage (and thus curl) is to use higher molecular weight oligomeric acrylates (96).

The problem of increased viscosity can be compensated by heating the resin during processing, although this solution is not universally applicable.

2.9.4 *Thiol-ene and Thiol-yne Systems*

Shrinkage is not the only problem from which acrylates suffer. Notably, propagating carbon radicals are inhibited by molecular oxygen dissolved in the resin (97). The problem is further exacerbated in open vat SLA setups where the curing surface is in constant contact with ambient air. Traditional additives for mitigating oxygen inhibition, such as tertiary amines, retard cationic polymerization and are not appropriate for mixed epoxy/acrylate resins. So, fast-curing alternatives to acrylates are to be considered.

One of the first alternate monomer systems to be investigated for SLA was based on thiol-ene chemistry (98,99). In this case, the *ene* component was actually a dinorbornene, which was formed by Diels-Alder reaction of a diacrylate (various diol acrylates, including hexanediol diacrylate) with cyclopentadiene. In equimolar ratio with a polythiol (pentaerythritol tetramercaptopropionate, cf. Figure 2.6, is cited in the example), the formulation cures with a much lower radiation dosage in comparison to DGEBA DA ($2\ mJ\ cm^{-2}$ vs. $13\ mJ\ cm^{-2}$). The authors reported the poor response of DGEBA DA on oxygen inhibition (100).

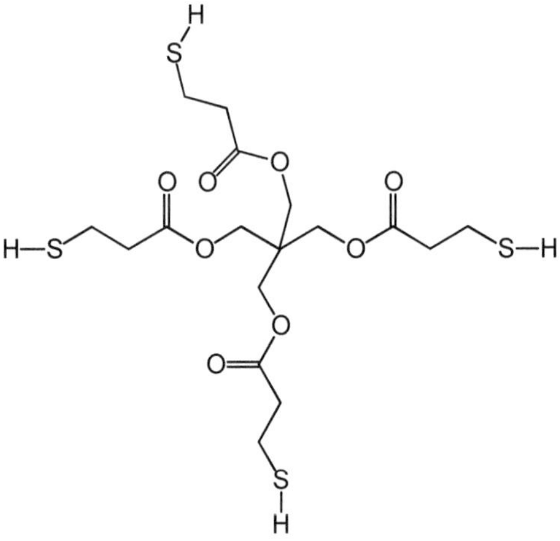

Figure 2.6 Pentaerythritol tetramercaptopropionate.

Thiols can alleviate oxygen inhibition by donating a hydrogen atom to a formed peroxyl radical, in the process providing a reactive thiyl radical. Other additives that have been tested to reduce oxygen inhibition in SLA include triphenylphosphine (76), where the authors specify resins with E_c values below 1 $mJ\,cm^{-2}$.

For SLA, the critical exposure E_c to cause gelation as measured in $mJ\,cm^{-2}$ can be defined as (73):

$$E_c = E_0 \times exp\left(-\frac{C_d}{D_p}\right) \tag{2.2}$$

In Eq. 2.2, E_0 is the dose at the surface, C_d is the curing depth, and D_p is the penetration depth.

Although thiols and phosphines can both improve in-air photocuring, they tend not to remain stable for extended times in formulations with acrylates. However, in the case of thiols, storage stability can be significantly improved by proper use of a buffer with a radical inhibitor (101).

The effect of various stabilizer systems on low-cytotoxic thiol/vinyl carbonate formulations were evaluated with the aim of inhibiting premature dark polymerization reactions (102). The addition of hydroxybenzene-based radical scavengers such as pyrogallol, hydroquinone monomethyl ether, or butylated hydroxytoluene (BHT) resulted in a significantly decelerated increase in viscosity compared to formulations without stabilizers during a defined storage period at 50°C. The usage of bicomponent stabilization systems in appropriate concentrations further reduced the viscosity increase for nonpigmented formulations. The most effective heterosynergistic stabilization system based on pyrogallol and diisoctyl phosphonic acid showed only a minor influence on the polymerization performance (102,103).

The compounds are shown in Figure 2.7.

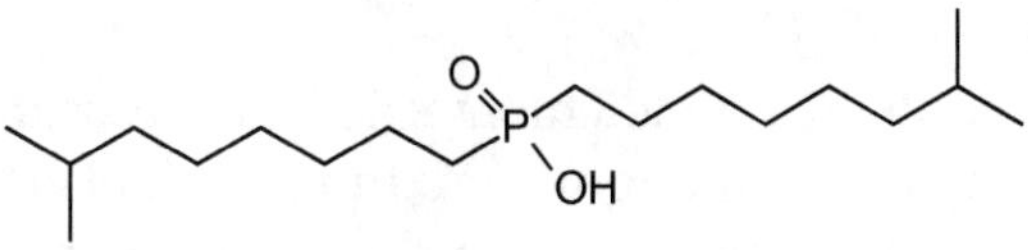

Diisoctyl phosphonic acid

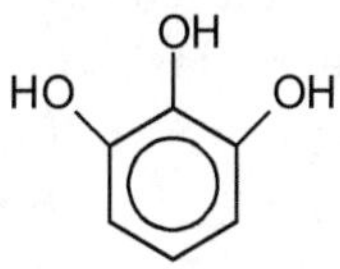

Pyrogallol

Figure 2.7 Heterosynergistic stabilization system.

While improved in-air curing is generally desirable for stereolithography, it does exclude the use of thiol-ene resins for continuous liquid interface production (CLIP), where oxygen inhibition is needed to prevent adhesion to the bottom of the vat (104).

2.9.4.1 *Spiroacetal Thiol-ene Systems*

A stable, tough resin formulation has been reported, which incorporates spiroacetal molecules into the polymer backbone and displays widely varying and tunable mechanical properties (105).

The materials used for synthesis are shown in Table 2.3.

Table 2.3 Materials used for synthesis of spiroacetal thiol-ene (105).

Compound
1,6-Hexanedithiol
Trimethylol propane tris(3-mercaptopropionate)
2,2-Dimethoxy-2-phenyl-acetophenone
Diphenyl(2,4,6-trimethylbenzoyl)phosphine oxide
3,9-Divinyl-2,4,8,10-tetraoxaspiro[5.5]undecane
Pyrogallol
Diisooctylphosphinic acid
2,2-Dimethoxy-2-phenyl-acetophenone

The thiol-ene resins were synthesized according to a previously described method (106).

Each resin contained a stoichiometric balance of thiol and vinyl functional groups in the monomer mixture, while also holding crosslinker, trimethylol propane tris(3-mercaptopropionate) concentration constant such that 7.5% of all thiol functional groups belong to trimethylol propane tris(3-mercaptopropionate). After mixing all monomers together in a clean glass vial, the mixture was heated to 75°C to expedite the mixing of crystalline 3,9-divinyl-2,4,8,10-tetraoxaspiro[5.5]undecane monomers with the liquid thiols (105).

2.9.5 *Epoxides*

Epoxides are one of the most commonly used classes of monomers for photo-based additive manufacturing. One reason is that epoxides undergo significantly less shrinkage (2%–3% vol) than acrylates during photocrosslinking (107). This fact can be explained by the ring opening reaction of the epoxide group. Another reason is the generally good mechanical properties of the resultant polymers.

The most commonly used epoxide monomers for SLA include diglycidyl ether derivatives of DGEBA, 3,4-epoxycyclohexylmethyl-3,4-epoxycyclohexanecarboxylate (ECC), cf. Figure 2.8, and epoxides of aliphatic alcohols such as trimethyloyl propane (93).

The reactivity of an epoxide monomer is dependent on molecular structure, where cycloaliphatic epoxides with high double ring strain like VCDE crosslink most rapidly. Epoxy monomers with

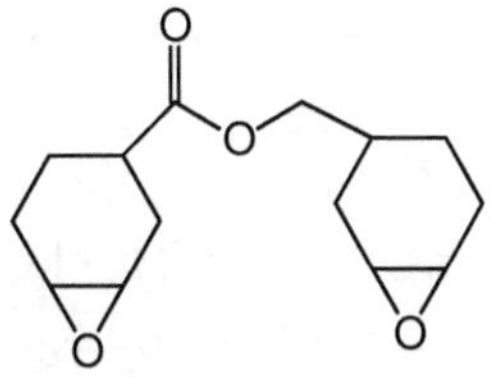

Figure 2.8 3,4-Epoxycyclohexylmethyl-3,4-epoxycyclohexanecarboxylate.

nucleophilic groups including ester moieties, which may be protonated, have reduced reactivity. Thus, photocrosslinking of ECC is about 10 times slower than that of other cyclohexene-derived epoxides without nucleophilic groups. Complexation of the ester group of ECC both intra- and intermolecularly with oxiranium intermediates can retard the desired reaction (108).

Ether groups such as those found in epichlorohydrin-derived epoxides (e.g., DGEBA) can also form bidentate or multidentate proton coordination, which explains the lower reactivity of these monomers (109).

An increase in the melting point and in the enthalpy of melting of polyamides can be achieved by a water treatment. This method makes it possible to increase the melting point T_m and the enthalpy of melting of polyamides without appreciably modifying the crystallization temperature T_c, also called the solidification temperature. These properties of polyamides are useful in many applications and in particular in the technology of polyamide powder sintering by melting using radiation such as, for example, laser sintering (109).

2.10 Visible Light-Curable and Visible Wavelength-Transparent Resin

An ultraviolet or visible light source is critical for SLA printing technology. UV light can be used to manufacture 3D objects in SLA, but there are significant occupational safety and health issues, particularly for the eyes (110). These issues prevent the widespread use of SLA at home or in the office. Through the use of visible light, the safety and health issues can largely be solved, but only non-transparent 3D objects can be manufactured, which prevents the application of 3D printing to the production of various common transparent consumer products.

For these reasons, a polymeric resin for SLA three-dimensional printing has been developed (110). The key of the research was to identify the photoinitiator diphenyl(2,4,6-trimethylbenzoyl)phosphine oxide (DPTBP), cf. Figure 2.9. DPTBP was originally designed as a UV photoinitiator, but it was found that visible light irradiation is sufficient to split DPTBP and generate radicals due to its slight visible light absorption up to 420 *nm*.

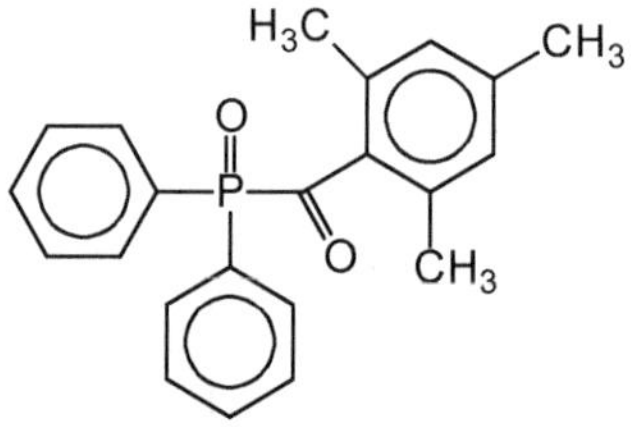

Figure 2.9 Diphenyl(2,4,6-trimethylbenzoyl)phosphine oxide.

The cured resin displays a high transparency and beautiful transparent colors by the incorporation of various dyes. In addition, its mechanical properties are superior to those of commercial resins (Arario 410) and photoinitiators (Irgacure 2959) (110).

2.11 Poly(ether ether ketone)

The additive layer manufacturing of poly(ether ether ketone) (PEEK) by means of FDM has been investigated. PEEK is a high-performance polymer with outstanding mechanical properties, high thermal stability and chemical resistance. It is suitable for space applications (111). However, due to its semicrystalline nature it is difficult to process. Moreover, only very few FDM printers suitable for PEEK are currently available on the market.

PEEK powder was extruded to obtain 1.75±0.05 *mm* diameter filaments using a single-screw extruder (Filabot, USA, screw speed 30 *rpm* and a temperature of 340°C). Tolerance of the diameter of the filament is a critical point as PEEK experiences high thermal shrinkage. Both powder and filament were dried for 24 *h* at 150°C and kept in a vacuum bag prior to use.

All filaments were successively processed with an INDMATEC GmbH FDM printer to produce tensile test specimens according to ASTM D638 Type V (112). Tensile testing specimens were printed in two main directions: vertically, Z, and horizontally on building platform plane, XY.

The results of mechanical, thermal, microstructural, and morphological testing of FDM printed PEEK samples have been reported. Some of them were compared with that of the extruded filament prior to printing.

The results evidence the effect of the process on the printed parts in terms of thermal and mechanical properties, including fracture mechanism. Moreover, the impact of printing parameters, such as infill and filament deposition pattern, on the final mechanical performance is evidenced too, as it is linked to the resisting cross section (111).

2.12 Lasers

The basics of lasers, optics and materials used for 3D printing and manufacturing have been described in a monograph (113). Also, several case studies have been included.

Furthermore, the fundamentals of light-matter interaction have been discussed with illustrative examples. Electromagnetic radiation and the three types of phenomena that occur when light interacts with matter have been explained. Then, the working principle of lasers and their characteristics were discussed. Collimation optics, polarization and different, but related, optical components which form the basic optics for laser-assisted manufacturing and 3D printing, have been detailed.

2.13 Ultra-High Molecular Weight PE

The development of customized prosthesis for total joint arthroplasty and total hip arthroplasty was elucidated (114). SLS as additive manufacturing could enable small-scale fabrication of customized ultra-high molecular weight poly(ethylene) (UHMWPE) components. However, the processes for SLS of UHMWPE needed to be improved.

Improving the preheating system of the SLS fabricating equipment was assessed. Then cuboids with the same size were fabricated along with cuboids of the same volume and different size to study the warpage, demonstrating the effect of the value and uniformity of the preheating temperature on component fabrication.

Warpage, density and tensile properties were investigated from the perspective of energy input density. Finally, complicated industrial parts could be effectively produced by using optimized technological parameters.

The results of the study indicated that components can be fabricated effectively after the optimization of the SLS technological parameters, i.e., the preheating temperature, the laser power, the scanning interval, and the scanning speed. The resulting warpage was found to be less than 0.1 *mm* along with a density of 83.25 and tensile strength up to 14.1 *MPa*. UHMWPE sample parts with good appearance and strength are obtained after ascertaining the effect of each factor on the fabrication of the sample parts. In summary, it was found to be very challenging to fabricate UHMWPE sample parts by SLS (114).

2.14 Production of PP Polymer Powders

New approaches towards the production of polymer powders for selective laser beam melting of polymers have been shown (115).

A new field of application of powder-based additive manufacturing methods is selective laser beam melting of polymers. However, there is only a limited choice of materials that show a good processability.

Hitherto, only PA-based materials were available as optimized powders for laser beam melting. Two innovative methods for production of spherical polymer microparticles for laser beam melting of polymers have been presented (115):

1. Wet grinding with subsequent rounding in a heated downer reactor, and
2. Melt emulsification.

Also, the possibilities of dry coating to tailor the particle properties have been illustrated. The influence of particle and powder properties like bulk density, shape or flowability on the laser beam melting processability and properties of devices obtained from the powders were outlined.

The first approach, stirred media milling and rounding, is applicable for a variety of polymers with a quite high breakage elongation. The second approach, i.e., melt emulsification, allows for spherical polymer microparticles in a single process step.

The dependency of powder properties on process parameters have been outlined exemplarily for the production of spherical poly-(propylene) (PP) particles. These PP powders with a good flowability could be obtained after spray drying of the product suspensions. The powders produced by the two methods described above were characterized with respect to tensile strength, packing density and layer formation. Moreover, the processabilty of the PP powder could be proven by building single layers (115).

2.15 Acrylate-Based Compositions

2.15.1 Dimensionally Stable Acrylic Alloys

An acrylic alloy composition has been described that can be 3D printed by a material extrusion additive manufacturing process. The acrylic filament has a very uniform diameter useful in the extrusion additive manufacturing process.

The acrylic alloy composition is an alloy of an acrylic polymer, and a low melt viscosity polymer, such as PLA. The alloy may optionally be impact modified, preferably with hard core-shell impact modifiers.

The acrylic polymers include polymers, copolymers, and terpolymers formed from alkyl methacrylate and alkyl acrylate monomers, and mixtures. The alkyl methacrylate monomer is preferably methyl methacrylate, which may make up from 50% to 100% of the monomer mixture, 0% to 50% of other acrylate and methacrylate monomers or other ethylenically unsaturated monomers, including styrene, α-methyl styrene, acrylonitrile, and crosslinkers at low levels, may also be present in the monomer mixture. Other methacrylate and acrylate monomers useful in the monomer mixture are collected in Table 2.4.

Table 2.4 Methacrylate and acrylate monomers (116).

Acrylate monomer	Methacrylate monomer
Methyl acrylate	Methyl methacrylate
Ethyl acrylate	Ethyl methacrylate
Butyl acrylate	Butyl methacrylate
i-Octyl acrylate	*i*-Octyl methacrylate
Lauryl acrylate	Lauryl methacrylate
Stearyl acrylate	Stearyl methacrylate
Isobornyl acrylate	Isobornyl methacrylate
Methoxy ethyl acrylate	Methoxy ethyl methacrylate
2-Ethoxy ethyl acrylate	2-Ethoxy ethyl methacrylate,
Dimethylamino ethyl acrylate	Dimethylamino ethyl methacrylate

The acrylic composition can be extruded into a filament for use in a material extrusion additive process. The filament can be a single strand of the acrylic composition, or can be in the form of a coextruded multi-phase filament.

The filament can have a middle layer composition of the acrylic alloy, surrounded by a sheath containing a different acrylic composition, for example, containing a special effects additive, or visa versa. The filament formed from the acrylic composition of the invention has only a little shrinkage or warpage, forming a filament that is very uniform in diameter. The low variance in the diameter of the filament is a key to its use in the material extrusion additive manufacturing process, as the calculations on the feed rate, and thereby the resulting density of the printed article, are based on calculations assuming a constant filament diameter (116).

2.15.2 *Oligoester Acrylates*

Liquid photopolymerizable pasty compositions suitable for 3D printing have been presented (117). Such a composition can be used in conjunction with a hand-held, self-contained 3D printing device, such as a 3D printing pen.

The oligoester acrylate is a dimethacrylic ester of triethylene glycol, which is a nontoxic homogeneous transparent liquid soluble in organic solvents and insoluble in water.

Liquid poly(ethylene glycol) (PEG) is a liquid, which is non-volatile at room temperature and is selected from the group consisting of PEGs with molecular weights in the range of 200 *Dalton* to 600 *Dalton*. The non-liquid PEG has a molecular weight in the range of 600 *Dalton* to 6,000 *Dalton*.

Catalysts for the synthesis of low molecular weight PEGs are alkali NaOH and KOH, as well as ash Na_2CO_3. The molecular weight of the polymer in this case is defined by the initial ethylene glycol and ethylene oxide as the polymerization proceeds without opening of the circuit (117).

The composition contains 60% to 80% of oligoester acrylate, 10% to 30% of a liquid PEG, 7% to 9% of a non-liquid PEG, and 0.1% to 1% of a system of photopolymerization initiators.

The pasty basic polymer can be obtained by mixing the components in the aforementioned proportions, stirring the obtained solution for 1 to 6 *h* in a magnetic stirrer, and then subjecting a third of the solution to UV radiation until reaching a solid state. The obtained solid material is crushed in a mill until it becomes a powder. The remaining third of the solution is subjected to UV curing under conditions of constant stirring until a uniform jelly is formed. The obtained photopolymerized jelly is mixed with the powder unit and a uniform mass can be produced (117).

2.16 Standards

Standards, quality control, and measurement sciences in 3D printing and additive manufacturing have been described in a monograph (118). Some standards are listed in Table 2.5.

Table 2.5 Standards for 3D printing.

Number	Title	Reference
ASTM E2544-11a	Standard Terminology for Three-Dimensional (3D) Imaging Systems	(119)
ASTM E2641-09	Standard Practice for Best Practices for Safe Application of 3D Imaging Technology	(120)
ISO/IEC 14496-27	Printing material and 3D graphics coding for browsers conformance	(121)

2.16.1 *Biomedical Applications*

Three-dimensional printing or additive manufacturing of medical devices and scaffolds for tissue engineering, regenerative medicine, *ex-vivo* tissues and drug delivery has become of intense interest in recent years (122).

A few medical devices, namely ZipDose®, pharmacoprinting, powder bed fusion, HPAM™, bioprinter and inkjet printer received FDA clearance while several biomedical applications are being developed.

The influence of the type of the additive manufacturing method and the process parameters on the surface topography, geometrical features, mechanical properties, biocompatibility, *in-vitro* and *in-vivo* performance of diverse orthopedic applications have been reviewed.

Also, attempts have been made to identify gaps, suggest ideas for future developments, and to emphasize the need for standardization (122).

2.16.2 *Color*

A broad view has been provided regarding the standardization efforts of color quality evaluation of color 3D printing techniques (123).

The processes and color properties of most color 3D printing techniques with specific devices and applications have been reviewed.

The design, methodology, and approaches of six color 3D printing techniques, including plastic-based, paper-based, powder-based, organism-based, food-based and metal-based color 3D printing, have been introduced and illustrated with colorization principles and forming features. Also, for printed 3D color objects, the literature about color measurement, color specification and color reproduction has been described and analyzed (123).

These six color 3D printing techniques vary from colorization materials to processes. However, their similarity has its basis in the general subtractive color theory, as recommended and standardized by the ICC (International Color Consortium) and the CIE (Commission Internationale de L'Eclairage) (124).

Four color 3D printing techniques, including plastic-based, paper-based, powder-based, and food-based color 3D printing, show a great affinity toward standardization of color quality evaluation, while their colorization principles indicate that this is difficult using only a single standard frame.

However, it is possible to develop a completed color quality evaluation standard for color 3D printing based on approaches in color 2D printing when color measurement method and devices are standardized together (123).

2.17 Particle-Free Emulsions

The development of yield-stress fluids that can be used in post-printing transformation processes is required by 3D printing (125). There is only a limited number of yield-stress fluids currently available with the desired rheological properties for building structures with small filaments of 100 μm with a high shape-retention.

A printing-centric approach for 3D printing particle-free silicone oil-in-water emulsions with a polymer additive, poly(ethylene oxide), has been presented (125).

This particular material structure and formulation is used to build 3D structure and to pattern at filament diameters below that of any other known material in this class. Increasing the molecular weight of poly(ethylene oxide) drastically increases the extensibility of the material without significantly affecting shear flow properties (shear yield stress and linear viscoelastic moduli). Higher extensibility of the emulsion correlates to the ability of filaments to span relatively large gaps (greater than 6 mm) when extruded at large tip diameters (330 μm) and the ability to extrude filaments at high print rates (20 $mm\,s^{-1}$. 3D printed structures with these extensible particle-free emulsions undergo a post-printing transformation, which converts them into elastomers. These elastomers can buckle and recover from extreme compressive strain with no permanent deformation, a characteristic not native to the emulsion (125).

2.18 Shape Memory Polymers

There are recent monographs dealing with the issues of shape memory polymers (126, 127).

Shape memory polymers (SMPs) are a class of programmable stimuli-responsive shape changing polymers. They are attracting increasing attention from the standpoint of both fundamental research and technological innovations. The progress in new shape memory enabling mechanisms and triggering methods, variations in shape memory forms, new shape memory behavior, and novel fabrication methods have been reviewed (128). Also, some research activities on 4D bioprinting have been detailed (129).

The SMPs are defined by the ability to recover their permanent shapes from one or sometimes multiple programmed temporary shapes when a proper stimulus is applied, such as temperature, magnetic field, light and moisture, and others.

The feasibility of using 3D printing technique has been demonstrated to create functional graded SMPs with both spontaneous and sequential shape recovery abilities (130).

The manufacturing method involves direct 3D printing from a CAD file that specifies the details of the material configuration and property distribution, which provides considerable freedom for design and operation convenience during the creation of functionally graded SMPs. An epoxy-based UV curable SMP was used as polymeric material, whose shape recovery is thermally triggered. By controlling the composition of the printed materials, the glass transition temperatures of the SMPs can be controlled to create a functional gradient. By properly specifying the material properties in different sections, it was demonstrated that the deformed SMP component can successfully return back to the original configuration in a predefined sequence, while the shape recovery in a SMP component without any property gradient will either be impeded or even fail in the middle. By using this method, even complex 3D solids can be created with arbitrarily defined material distributions, which provide a potential route for precisely controlling the shape recovery profile and enabling the fabrication of devices with unprecedented multifunctional performance.

The created SMP components, with properly assigned spatial variation of the thermodynamical property distribution, react rapidly to a thermal stimulus, and return to a specified configuration in a precisely controlled shape changing sequence. The use of the 3D printing technique enables a manufacturing routine with the merits of easy implementation, large design freedom, and high printing resolution, which promises to advance immediate engineering applications for low-cost, rapid, mass production (130).

2.18.1 *Synthesis with Stereolithography*

A dual-component mechanism can be used to synthesize SMP for a stereolithography process (131). The materials can be readily polymerized at a rapid laser curing speed with high accuracy.

The method uses *tert*-butyl acrylate as monomer and di(ethylene glycol) diacrylate as crosslinking agent. As UV photoinitiator, phenylbis(2,4,6-trimethylbenzoyl)phosphine oxide was used.

The material can achieve a high curing rate and precise printing that are highly desired for the SLA process. The mechanical strength of the printed parts is comparable statistically to industrial SMP, and shape memory tests showed excellent shape memory performance with higher durability of more then 20 shape memory cycles as compared to current 4D printed parts.

So it was demonstrated that the use of SLA technology enables the fabrication of responsive components with high performance, which also significantly advances the 3D printing technology for more robust applications (131).

2.18.2 *Flexible Electronics*

The formation of 3D objects composed of shape memory polymers for flexible electronics has been described (132).

A layer-by-layer photopolymerization process of methacrylated semicrystalline molten macromonomers by a 3D digital light processing printer enables the rapid fabrication of complex objects and imparts shape memory functionality for electrical circuits.

The shape memory thermosets were based on poly(caprolactone) (PCL). PCL is solid at room temperature and has a melting temperature between 43°C and 60°C depending on its molecular weight. To prepare the resin, methacrylate groups were covalently linked to the chain ends of the PCL macrodiol using a simple alcohol-isocyanate reaction. In addition, the PCL formulations contained 4% 2,4,6-trimethylbenzoyl diphenyl phosphine oxide as photoinitiator.

It is possible to fabricate complex shape memory structures with a viscous melt of 30 *Pa s* using a commercial SLA printer and a customized heated resin bath of the system setup. The 3D printing was performed by layer-by-layer UV curing at the bottom of the reservoir, curtailing the deleterious effect of molecular oxygen inhibition (133).

The build times were dependent on the size of the models and the layer thicknesses. For example, it took 44 *min* to print a 1 cm^3 cube with a 100 μm layer thickness.

The optimized printing parameters are shown in Table 2.6.

Table 2.6 Optimized printing parameters for the Asiga Pico Plus 39 printer (132).

Build Parameter	Value
Slice thickness	0.1 *mm*
Burn-in layers	1 layer
Separation distance	15.0 *mm*
Separation velocity	5.0 mms^{-1}
Approach velocity	5.0 mms^{-1}
Slide velocity	10.0 mms^{-1}
Slides per layer	1
Exposure (burning time)	12.0 *s*
Exposure time	8.0 *s*
Wait time (after slide)	3.0 *s*
Wait time (after exposure)	0 *s*
Wait time (after separation)	3.0 *s*
Wait time (after approach)	2.0 *s*
LED wavelength	385 *nm*
Pixel size	39 μm
Light intensity	17.5 *mV*
Water bath temperature (heat source)	90 °C

The drawback of slow print speeds in hindering adoption of additive manufacturing processes is well known but recently a new SLA technique was reported with print speeds up to two orders of magnitude faster than the resin agnostic (104).

The X-Y axis resolution of the printer used in the study is 39

μm, whereas on the layer thickness (Z-axis), it can be as low as 1 μm. In the experiments, a layer thickness of 100 μm was used. The unreacted macromonomer was partially removed from the voids of the printed objects by immersion in warm isopropanol using a sonication procedure. A final curing was performed by additional UV exposure for 30 s (132).

2.18.2.1 Electro-responsive Nanocomposites

A simple approach to 3D printing of carbon-black-based shape memory polymer nanocomposites with toughness improving capabilities during programming stage using electrical stimulus has been reported (134).

Conductive shape memory polymer nanocomposites, consisting of commercial SMP filled with conductive carbon black nanoparticles, were fabricated using a solvent casting method and a single-screw extrusion process. Subsequently, a material extrusion technique was used for 3D printing of Type IV dog bone shaped specimens for testing tensile strength and electrical stimulation.

It was found that SMP/carbon black electrical conductivity can be tuned by the filler fraction. In addition, electrical current passing through SMP/carbon black nanocomposites causes temperature increments and changes in the material strength. Temperature profiles at various electrical current levels were reported.

It was observed that conductive SMP/carbon black specimens responded to electrical current stimulus by increasing their toughness four times higher than with no current applied during tensile testing (134).

2.18.2.2 Flexible Shape Changing Actuators

The fabrication of flexible shape changing actuators by means of 3D printing has become an exciting research area, which has been widely used in our daily lives and is expected to play more important roles in soft robotics, biomedical devices and other high-tech areas (135). However, the development of such 3D printed shape changing actuators is limited due to the lack of 3D printing functional materials and insufficient response sensitivity of the actuators.

The 3D printing of photo-responsive shape changing composites has been demonstrated (135). This is based on PLA and multi-walled carbon nanotubes on paper substrates with a FDM printing technology for the construction of flexible photothermal-responsive shape changing actuators. The introduction of multiwalled carbon nanotubes into a PLA matrix results in the enhancement of the processability of such composites.

Multiwalled carbon nanotubes could be dispersed homogeneously in the PLA matrix. Furthermore, the multiwalled carbon nanotube-PLA composites exhibit excellent photothermal effects and sensitivity under near-infrared irradiation. The temperature of the composite increases up to the glass transition temperature of PLA after 1 *s* irradiation and also is close to the melting temperature of PLA after an irradiation time of 15 *s*.

Thus, paper-based bilayer semicircular actuators that possess phototriggered shape changing properties were fabricated via 3D printing of the multiwalled carbon nanotube-PLA composite on paper, which deform under near-infrared irradiation and recover their original shape once the light source is switched off. This printing strategy for flexible paper-based actuators could provide tremendous opportunities for the design and fabrication of biomimetic photothermal actuators and soft robotics (135).

2.18.3 Magnetically Responsive Shape Memory Polymer

A 3D printing technique has been used for the fabrication of functionally graded magnetic SMPs to create high-resolution multimaterial shape memory architectures (136).

This approach was applied using a copolymer network of photocurable methacrylate using high projection stereolithography. Carbonyl iron particles were physically embedded into a polymer matrix to add a magnetic functionality to the SMPs. The glass transition characteristics and shape memory effect were also investigated by varying the composition of the SMP.

The microstructured, lightweight SMPs showed interesting shape memory behaviors, as observed in a hot environment. The almost perfect strain recovery rate of poly(ethylene glycol) dimethacrylate was measured to be 99.95% using a tension set bar. The results of dynamic mechanical analysis and thermogravimetric analysis revealed an increment in the thermal conductivity after embedding the carbonyl iron particles. Furthermore, the results of dynamic mechanical analysis, differential scanning calorimetry, and scanning electron microscopy revealed a close interaction between the particles and matrix. X-ray diffraction was used to characterize the iron particles and the structure of the polymer.

The so obtained results, along with the electrical and magnetic tests, strongly support the remote controllability of the material properties of the functionally graded magnetic SMPs for a broad range of temperature and/or magnetically responsive material applications by using eddy current heating and/or magnetorheological polymeric effects (136).

2.18.4 *Sequential Self-Folding Structures*

Sequential self-folding structures could be realized by the thermal activation of spatially variable patterns that were 3D printed with digital shape memory polymers, which are digital materials with different shape memory behaviors (137).

Two base model materials were used, i.e., VeroWhite and TangoBlack. VeroWhite is a rigid plastic at room temperature polymerized with ink containing isobornyl acrylate, acrylic monomer, urethane acrylate, epoxy acrylate, acrylic monomer, acrylic oligomer, and photoinitiators. TangoBlack is a rubbery material at room temperature polymerized by monomers containing urethane acrylate oligomer, exo-1,7,7-trimethylbicyclo[2.2.1]hept-2-yl acrylate, methacrylate oligomer, poly(urethane) resin and photoinitiators. The tested materials consisted of varying compositions of these two materials that lead to different thermomechanical properties. As printing method, jet spraying before UV curing was used.

The time-dependent behavior of each polymer allows the temporal sequencing of activation when the structure is subjected to a uniform temperature. This was demonstrated via a series of 3D printed structures that respond rapidly to a thermal stimulus, and self-fold to specified shapes in controlled shape changing sequences.

The spatial and temporal nature of the self-folding structures are in good agreement with finite element simulations. A simplified reduced-order model has also been developed to rapidly and accurately describe the self-folding physics.

An important aspect of self-folding is the management of self-collisions, where different portions of the folding structure contact and then block further folding. A metric has been developed to predict collisions and was used together with the reduced-order model to design self-folding structures that lock themselves into stable desired configurations (137).

2.18.4.1 *Hydrogels*

A reversible shape changing component design concept has been described (138). This is enabled by 3D printing of two stimuli-responsive polymers: Shape memory polymers and hydrogels.

This approach uses the swelling of a hydrogel as the driving force for the shape change, and the temperature-dependent modulus of a shape memory polymer to regulate the time of such shape change. Controlling the temperature and aqueous environment allows switching between two stable configurations: The structures are relatively stiff and can carry load in each, without any mechanical loading and unloading.

The digital materials Grey60 and TangoBlack were used as materials. Grey60 is a digital material with the glass transition temperature of ~ 48°C, which can be used as the SMP when the temperature is changed between 0°C and 60°C.

Specific shape changing scenarios, e.g., based on bending or twisting in prescribed directions, are enabled via the controlled interplay between the active materials and the 3D printed architectures. The physical phenomena are complex and nonintuitive, and so to help understand the interplay of geometric, material, and environmental stimuli parameters we develop 3D nonlinear finite element models (138).

Hydrogel Actuators. Porous structures have emerged as a breakthrough of shape-morphing hydrogels to achieve a rapid response (139, 140). However, these porous actuators generally suffer from a lack of complexity and diversity in obtained 3D shapes.

A simple and versatile strategy has been developed to generate shape-morphing hydrogels with both fast deformation and enhanced designability in 3D shapes by combining two technologies: electrospinning and 3D printing.

Elaborate patterns could be printed on mesostructured stimuli-responsive electrospun membranes, modulating in-plane and interlayer internal stresses induced by a swelling/shrinkage mismatch, and thus guiding morphing behaviors of electrospun membranes to adapt to changes of the environment.

Using these methods, a series of fast deformed hydrogel actuators could be constructed with various distinctive responsive behaviors, including reversible/irreversible formations of 3D structures, folding of 3D tubes, and formations of 3D structures with multi-low-energy states.

Poly(*N*-isopropyl acrylamide) was chosen as the model system in this study; however, it was stated that other stimuli-responsive hydrogels can also be used (140).

2.18.5 *Multi-shape Active Composites*

Recently an exciting new dimension to 3D printing technology appeared. After being printed, these active, often composite, materials can change their shape over time. This is referred to as 4D printing (141). The design and manufacture of active composites has been shown that can take multiple shapes, depending on the environmental temperature.

This can be achieved by 3D printing layered composite structures with multiple families of SMP fibers–digital SMPs–with different glass transition temperatures to control the transformation of the structure. After a simple single-step thermomechanical programming process, the fiber families can be sequentially activated to bend when the temperature is increased. By tuning the volume fraction of the fibers, the bending deformation can be controlled.

Two families of digital SMP fibers with different glass transition temperatures were embedded in the two layers, respectively, with prescribed volume fractions. The fibers (fiber 1: DM8530, glass transition temperatures ~ 57°C; fiber 2: DM9895, glass transition temperature ~ 38°C) have shape memory effects in the temperature range between ~ 20°C and ~ 70°C.

Samples were fabricated using an Objet 3D printer (Objet 260), which can mix two base materials to make digital materials with different mechanical properties. As materials, VeroWhite and TangoBlack were used. During the printing process, the printer can mix the two materials with different component ratios to achieve the material with the desired mechanical properties.

Due to the advantage of an easy fabrication process and the controllable multi-shape memory effect, the printed SMP composites have great potential in 4D printing applications (141).

2.18.6 *Radiation Sensitizers*

The mechanical properties of materials that are printed using fused filament fabrication 3D printers typically rely only on the adhesion among the melt of processed thermoplastic polymer strands (142). This dramatically limits the utility of fused filament fabrication systems for a host of manufacturing and consumer products and severely limits the toughness in 3D printed shape memory polymers.

In order to improve the interlayer adhesion in 3D printed parts, crosslinked units can be introduced among the polymer chains by exposing the 3D printed copolymer blends to ionizing radiation. This strengthens the parts and reduces anisotropy. A series of polymers blended with radiation sensitizers, such as trimethylol propane triacrylate and triallyl isocyanurate, cf. Figure 2.10, were prepared and irradiated by γ-rays. Several methods were employed to characterize the thermomechanical properties and the chemical structure of the prepared polymers.

Triallyl isocyanurate was shown to be a very effective radiation sensitizer for 3D printed sensitized PLA. It was further shown that crosslinks induced by radiation temperatures near the T_g of shape memory systems prominently enhanced the thermomechanical properties of the 3D printed polymers, as well as the solvent resistance (142).

2.18.7 *Shape Memory Alloy Actuating Wire*

A process for the embedding of shape memory alloy actuating wire within direct PolyJet 3D printed parts has been presented (143). A series of *Design for Embedding* considerations were shown for achieving successful and repeatable embedding results.

These considerations include guide channel design, design of shape converters for irregularly shaped elements, and design of

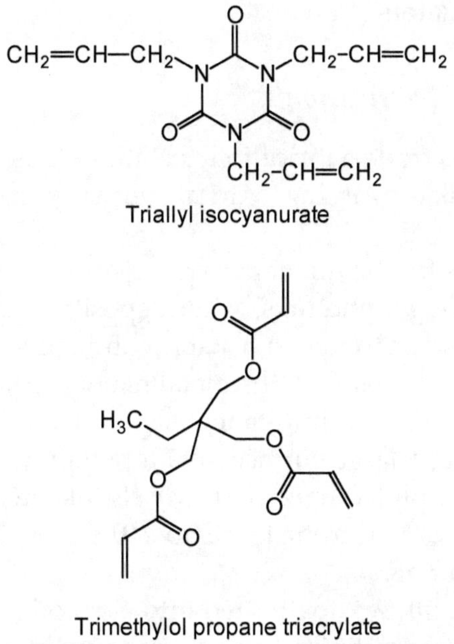

$CH_2=CH-CH_2$ $CH_2-CH=CH_2$

$CH_2-CH=CH_2$

Triallyl isocyanurate

H_3C

Trimethylol propane triacrylate

Figure 2.10 Radiation sensitizers.

wire fixation points. The embedding process was demonstrated with two case studies (143):

1. A simple compliant joint specimen with a straight shape memory alloy wire, and
2. An antagonistic joint design with spring-shaped shape memory alloys.

The process is characterized through an exploration of the potential for surface defects in the final specimens, as well as basic quantitative and qualitative evidence regarding the performance of the final embedded actuators (143).

2.18.8 Metal Electrode Fabrication

A simple 3D metal electrode fabrication technique has been reported (144). Here, inkjet printing onto a thermally contracting SMP substrate is used.

Inkjet printing allows for the direct patterning of structures from metal nanoparticle bearing liquid inks. After deposition, these inks require thermal curing steps to render a stable conductive film.

By printing onto a SMP substrate, the metal nanoparticle ink can be cured and substrate shrunk simultaneously to create 3D metal microstructures, forming a large surface area topology well suited for energy applications. Poly(styrene) (PS) SMP shrinkage was characterized in a laboratory oven from 150°C to 240°C, resulting in a size reduction of 1.97 to 2.58.

A silver nanoparticle ink was patterned into electrodes, shrunk, and the topology was characterized using scanning electron microscopy. Zinc-silver oxide microbatteries have been fabricated to demonstrate the 3D electrodes compared to planar references (144). The characterization was performed using a 10 M potassium hydroxide electrolyte solution doped with zinc oxide (57 $g\,l^{-1}$). After a 300 s oxidation at 3 V_{dc}, the 3D electrode battery demonstrated a 125% increased capacity in comparison to a reference cell.

Reference cells degraded with longer oxidations, but the 3D electrodes were fully oxidized for 4 h, and exhibited a capacity of 5.5 $mA\,h\,cm^{-2}$ with a stable metal performance (144).

The simple microbattery geometry used here validated the concept of using inkjet printed silver on shape memory polymer as a simple method to create 3D electrodes to boost the performance and stability over planar thin film electrodes. However, the electrolyte solution ultimately degraded the packaging materials, necessitating the development of alternatives (144).

2.18.9 4D Printing

2.18.9.1 Dynamic Jewelry and Fashionwear

A methacrylated semicrystalline polymer was used to print objects that exhibited a thermally triggered shape memory behavior. By exploring various molecular weights, it was found that a methacrylated PCL polymer with a number average molecular weight of 10,000 $gmol^{-1}$ exhibited the best thermal and mechanical behavior (145).

The effect of the addition of dye to the ink formulation in the photopolymerization and printing processes was evaluated. The ink was utilized for demonstrating the fabrication of dynamic jewelry and shoe accessories by digital light processing printing (145).

2.18.9.2 Smart Textiles

The creation of a new material that is capable of changing its shape when exposed to different stimuli and possesses the ability to be 3D printed can be a difficult and long process. Due to this strenuous process, the potential of a common fused deposition modeling material, PLA, for use in 4D printing was investigated and the concept of combining PLA with nylon fabric for the creation of smart textiles was explored. PLA possesses thermal shape memory behavior and maintains these abilities when combined with a nylon fabric that can be thermomechanically trained into temporary shapes and return to their permanent shapes when heated.

Two types of 4D printing were assessed. In the first test, the PLA material was the only material used during the process and 3D printed directly onto the print bed. In a second experiment, the PLA material was 3D printed onto the nylon fabric.

The nylon fabric that was used for the textile printing was Solid Power Mesh Fabric Nylon Spandex, made up of 90% nylon and 10% spandex. The nylon fabric was cut into 40 $mm \times 40$ mm squares and measured 0.26 mm in thickness. A double-sided tape was placed onto the print bed and the cut nylon fabric was placed onto the tape for better adhesion.

Creo Parametric 3.0 modeling software was used to create 3D models which were converted into *.STL files. The created *.STL files were finally uploaded to the 3D printing software (Simplify3D) and the parts were 3D printed onto the nylon fabric. The printing speed was set to 100 $mm\,s^{-1}$ and the nozzle temperature was 230°C for all the printing tests (146).

The designs were placed into a pool of water at 70°C for 60 *s* and compressed within the water. After the spline was compressed, it was removed from the pool and allowed to cool to room temperature, which caused the compressed spline to harden. The spline maintained its temporary compressed shape under T_g. However, the compressed spline quickly expanded back to its original shape once it was returned to the 70°C pool with water.

In summary, the concept of smart materials combined with nylon textiles displays the possibility of using smart textiles for encapsulation and controlled release in response to their surrounding environment. The nylon fabric in the experiments serves more as a structure and non-active material, while the PLA serves as the smart material (146).

2.18.9.3 *Ultrafast 4D Printing*

Ultrafast 4D printing in less than 30 *s* has been reported (147).

Instead of direct printing in 3D, 4D printing relies on introducing stresses into a printed 2D structure (137, 148, 149). Such a process overcomes the typical layer-by-layer printing limitation.

However, currently known 4D printing techniques rely on a localized control of anisotropic filler orientation in a polymer matrix to generate stress. As such, the stress can only be manipulated in a pixel-after-pixel basis via ink writing. Such a sequential process inherently limits its speed despite the fact that it omits the time-consuming multilayer buildup in the z dimension.

In contrast, materials and processes have been reported that enable ultrafast printing of multidimensional responsive polymers, including hydrogels and shape memory polymers (147). Such a method utilizes a brief digital light exposure on light-curable monomers, requiring neither the layer-by-layer process in the vertical dimension nor the sequential pixel manipulation in the planar dimensions.

For printing of 3D hydrogels, the monomer mixture consists of hydroxyethyl acrylate, 2-hydroxyethyl methacrylate, potassium 3-sulfopropylmethacrylate, poly(caprolactone) diacrylate as crosslinking agent, and Irgacure 819 as photoinitiator. A UV exposure time below 15 *s* yields hydrogels with moduli below the testing limit.

Besides hydrogels, monomer systems can be selected that allow the digital printing concept to be expanded to other types of stimuli-responsive polymers. Printing of wax-based shape memory polymers has been demonstrated. To accomplish such polymers hydrophobic lauryl acrylate was used as the monomer and 1,6-hexanediol diacrylate as crosslinking agent. This monomer mixture was similarly cured by the digitally controlled light. The resulted polymer network can be swollen by melted wax due to the long aliphatic chain of lauryl acrylate.

The light exposure time significantly affects the swelling ratio, with a maximum swelling ratio of about 1.7 at an exposure time of 4 *s*. Figure 2.11 shows the dependance of swelling ratio and light exposure time.

In summary, visible-light-triggered polymerization of commercial monomers defines digital stress distribution in a 2D polymer film. Releasing the stress after the printing converts the structure into a 3D structure. An additional dimension can be incorporated by choosing the printing precursors. The process overcomes the speed-limiting steps of typical 3D (4D) printing (147).

2.19 Water-Soluble Polymer

During an additive manufacturing process, at least one material, i.e., object material or modeling material, is deposited to produce the desired object. But frequently a second material, i.e., support material or supporting material, is used to provide support for specific areas of the object during building and assure adequate vertical placement of subsequent object layers (150). Both materials, i.e, modeling material and supporting material, might be initially liquid and are subsequently hardened to form the required layer shape.

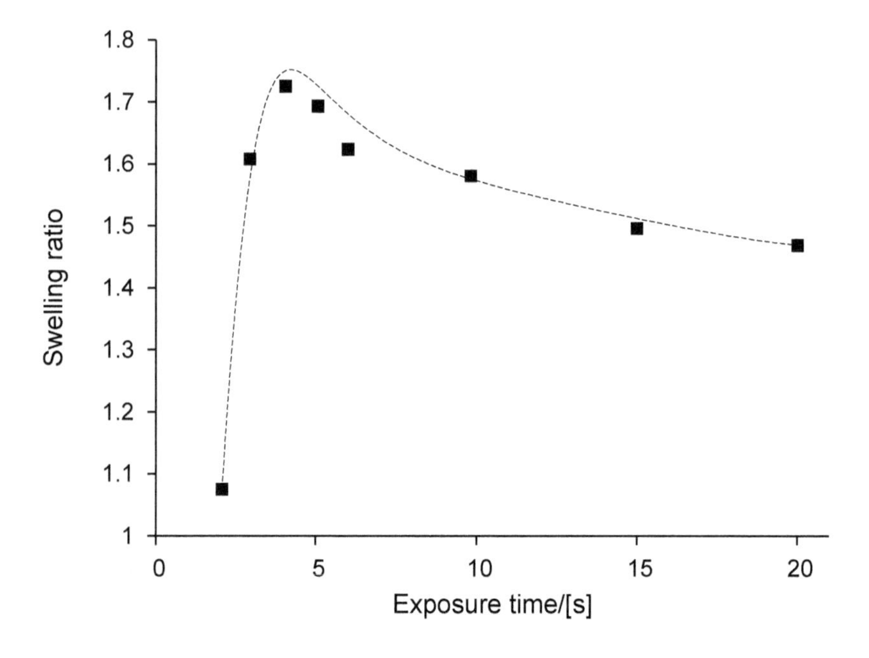

Figure 2.11 Dependance of swelling ratio and light exposure time (147).

The hardening process may be performed by a variety of methods such as UV curing, phase change, crystallization, drying, and others. In all cases, the support material is deposited in proximity to the modeling material, enabling the formation of complex object geometries and filling of object voids.

When using commercially available printheads, such as inkjet printing heads, the support material must have a relatively low viscosity of about 10–20 *cPs* at the working temperature, so it can be jetted. Furthermore, the support material should harden rapidly in order to allow building of subsequent layers. Additionally, the support material must have sufficient mechanical strength for holding the model material in place and low distortion for avoiding geometrical defects.

A water-soluble inkjet composition for 3D printing has been described (150). This material composition can be used as a support material.

The composition contains a glycol polymer and a low molecular weight polar substance, as well as appropriate surface active agents.

The polar substance is di-alcohol or a poly-alcohol. Examples for diols and polyols are shown in Table 2.7 and in Figure 2.12.

Table 2.7 Diols and polyols (150).

Diol	Polyol
1,8-Octanediol	1,2,4-Butanetriol
1,6-Octanediol	1,2,6-Hexanetriol
1,6-Hexanediol	1,2,3-Heptanetriol
1,3-Cyclopentanediol	2,6-Dimethyl-1,2,6-hexanetriol
trans-1,2-cyclooctanediol	1,2,3-Hexanetriol
3,6-Dithia-1,8-octanediol	1,2,3-Butanetriol
1,6-Hexanediol	3-Methyl-1,3,5-pentanetriol
1,7-Heptanediol	1,2,3-Cyclohexanetriol
1,9-Nonanediol	1,3,5-Cyclohexanetriol
1,2-Cyclohexanediol	3,7,11,15-Tetramethyl-1,2,3-hexadecanetriol
1,4-Cyclohexanediol	2-Hydroxymethyltetrahydropyran-3,4,5-triol
2,5-Dimethyl-3-hexyne-2,5-diol	1,1,1-Tris(hydroxymethyl)propane
2,5-Dimethyl-2,5-hexanediol	2,2-Diethyl-1,3-propanediol
	1,2,7,8-Octanetetrol

The surface active agent can be BYK-345, BYK-307, or BYK-347 (150). These compounds are silicone-based surfactants, polyether-modified siloxanes (151).

As an example, the material composition can contain about 70% 1,8-octanediol, about 30% PEG 1500, and about 0.5% BYK-345 (150).

It has been found that combining certain molecular weights of PEG (such as PEG 1500) with some low molecular weight polar substances, such as 1,8-octanediol, results in a material composition with lower viscosity at the elevated temperatures when compared with the same molecular weight of PEG alone. At the same time, this composition has a much faster solidification rate at the same lower temperature (150).

Comparative data for compositions with different glycol polymer, different low molecular weight di-alcohols or poly-alcohols are shown in Table 2.8.

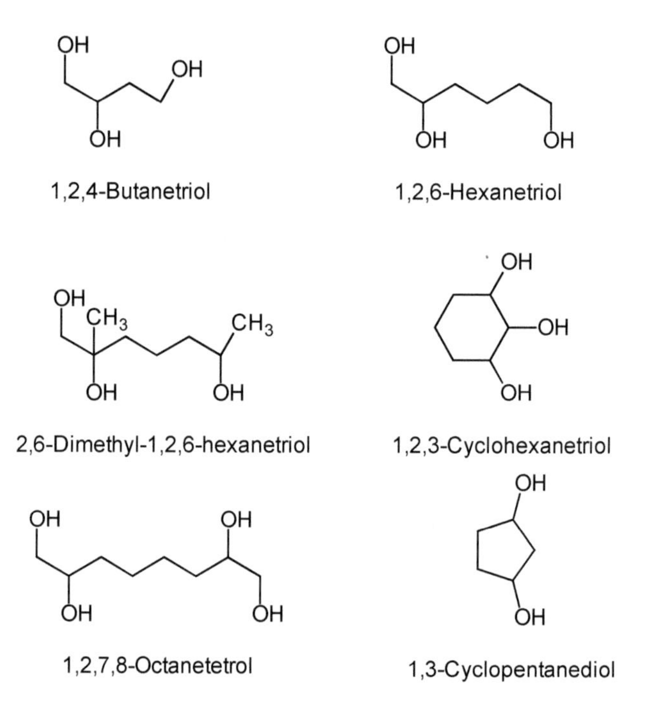

Figure 2.12 Diols and polyols.

Table 2.8 Comparative data for compositions (150).

Mixture	Melting point/[°C]	Dissolution time 40°C/[*min*]
1,10-Decanediol 100%	70	–
1,10-Decanediol 20%-PEG1500 80%	47	40
1,8-Octanediol 100%	61	70
1,8-Octanediol 50%-PEG1500 50%	50-55	11
1,8-Octanediol 75%-PEG1500 25%	56-58	15
1,8-Octanediol 50%-palmitic acid 50%	<50	–
1,8-Octanediol 50%-myristic acid 50%	<50	15
1,8-Octanediol 50%-octadecanol 50%	53	~300
1,8-Octanediol 33%-octadecanol	51	–
1,8-Octanediol 75%-hexadecanol 10%-PEG1500	52-55	25
25% 1,8-Octanediol 75%-hexadecanol 5%-PEG1500	55-57	15
25% Dimethyl hexanediol	86-90	10
Dimethyl hexanediol 50%-PEG1500 50%	70-80	12
Dimethyl hexanediol 75%-PEG1500 25%	70-85	15
Dimethyl hexanediol 75%-PEG4000 25%	80-87	15

2.20 Water-Washable Resin Formulations

One challenge encountered with stereolithography is the presence of uncured or partially cured resin on the surfaces of the printed 3D object. The uncured resin must be removed since partially cured resins render the object tacky and difficult to clean. Because most of the current resin formulations apply monomers or oligomers that are soluble only in certain organic solvents, the 3D printed objects need to be washed by rinsing or utilization of ultrasonication with organic solvents such as isopropanol or acetone.

The organic solvents are often volatile organic chemicals which are regulated by the Environmental Protection Agency (EPA) and Occupational Safety and Health Administration (OSHA). The use of organic solvents has several disadvantages (152):

1. The organic solvents are flammable and harmful to the human body and environment, making them less than ideal chemicals for use by ordinary users of 3D printing systems having little or no knowledge of chemistry,
2. The storage and disposal of organic solvents can be challenging,
3. Due to shrinkage during stereolithography, 3D printing can result in printing inaccuracies, such as shrunken size or even warped objects, as well as a high probability of failed printing, such as detachment between two adjacent thin layers, and
4. Organic solvents can add costs to printing the 3D objects.

Water-washable resin formulations have been presented that have a water-soluble oligomer, an at least partially water-soluble monomer, a photoinitiator and a light blocker, which also acts as an optical brightener. Formulations are detailed in Table 2.9.

The amount of water-soluble ingredients in the resin formulation is sufficient to allow the printed 3D objects using this formulation resin to be washed with water and leaving a dry surface. Instead of trimethylol propane triacrylate, other water-soluble oligomers and monomers may also be used. In addition, a water-soluble filler may be added (152).

Also, a difunctional epoxy monomer can be used in the formulation. The light curing process of the epoxy monomer starts with

Table 2.9 Water-washable resin formulations (152).

Formulation 1		
Material	Amount/[%]	Function
Miramer PE220	94.03	Oligomer
Trimethylol propane triacrylate	4.95	Monomer
Irgacure TPO	0.99	Photoinitiator
Benetex OB Plus	0.03	Blocker/Brightener
Formulation 2		
Material	Amount/[%]	Function
Miramer PE220	4.00	Oligomer
Trimethylol propane triacrylate	94.98	Monomer
Irgacure TPO	0.99	Photoinitiator
Benetex OB Plus	0.03	Blocker/Brightener

photon absorption by a photosensitizer after which the absorbed energy is transferred to the photoacid generator, starting the generation of acid, more particularly protons. The hardening of epoxy monomer is triggered by the protons generated by the photoacid generator.

Here, the curing of acrylate oligomer/monomer first forms the initial shape of the printed 3D objects and the epoxy in the resin gradually hardens the printed 3D objects because acrylate oligomer/monomer cures faster than the epoxy correspondents. In such formulations, the water-soluble ingredients may comprise 1% or greater of the water-washable resin formulation.

There are several benefits of using epoxy in a radiation-curable formulation. First, the epoxy monomer serves as a diluent having a low viscosity, which allows easier processing of the resin. Second, it is well known that the curing of epoxy monomer has lower shrinkage and therefore the shrinkage of the formulation is lower compared with the acrylate components only. Third, the inclusion of epoxy monomer improves the printed 3D objects' mechanical properties by increasing toughness and tensile strength. Fourth, the epoxy monomer also improves the finishing of the printing by post-curing because the curing of epoxy is not inhibited by oxygen and is relatively slow (152).

2.21 Extremely Viscous Materials

Heterogeneous materials used in biomedical, structural and electronics applications contain a high fraction of solids of more than 60% by volume (153). Also, they exhibit extremely high viscosities of more than 1000 *Pa s*, which hinders their 3D printing using conventional technologies.

It has been shown that by the use of high-amplitude ultrasonic vibrations within a nozzle imparts sufficient inertial forces to these materials to drastically reduce effective wall friction and flow stresses, enabling their 3D printing with moderate back pressures at high rates and with precise flow control.

This effect has been utilized to demonstrate the printing of a commercial polymer clay, an aluminum-polymer composite and a stiffened fondant with viscosities up to 14,000 *Pa s* with minimal residual porosity at rates comparable to thermoplastic extrusion.

This method can significantly extend the type of materials that can be 3D printed to produce functional parts without relying on special shear/thermal thinning formulations or solvents to lower viscosity of the plasticizing component. The high yield strength of the printed material also allows free-form 3D fabrication with minimal need for supports (153).

2.21.1 Tunable Ionic Control of Polymeric Films

A method has been reported to synthesize materials derived from highly viscous or even solid monomers in a simple, flexible fashion (154). These materials have the potential to be integrated into the printing process. Polymerizable ionic liquids have been employed due to the broad range of their properties available upon fine-tuning of the anion-cation pair and the high viscosity of the monomers.

The method consists of the deposition and polymerization of a polymerizable ionic liquid precursor, followed sequentially by quaternization and anion metathesis of the films. The fine control over the mechanical and superficial properties of inkjet printable polymeric films of neutral and cationic nature by postpolymerization reactions could be demonstrated.

A family of different polycationic materials has been generated by modification of crosslinked copolymers of butyl acrylate and vinyl imidazole with liquid solutions of functional reagents. The variation in the mechanical, thermal, and surface properties of the films demonstrates the success of this approach. The same concept has also been applied to a modified formulation designed for optimal inkjet printing (154).

2.22 Photopolymer Compositions

Among various 3D printing approaches, the photochemical approach is extremely attractive since objects can be produced via photopolymerization reactions of monomers/oligomers, which possesses environmental, economical, and production benefits (155).

Here, photoinitiators and monomers/oligomers are two of the most important components of photopolymers for 3D printing. Photoinitiators absorb the irradiation light of a 3D printer and initiate the photopolymerization reactions of the monomers/oligomers layer-by-layer to produce the designed 3D objects, while the type of monomers/oligomers can determine the final properties of the printed products. Commercial photoinitiators and monomers/oligomers applicable for 3D printing have been reviewed (155).

Recently developed photoinitiators and monomers/oligomers for 3D printing of polymer-based materials with various properties and their application in 3D bioprinting have also been demonstrated (155).

2.22.1 *Mechanical Properties of UV Curable Materials*

Considerable attention has been paid to the 3D printing technique with photopolymerization due to their high resolutions (156). Unfortunately, the 3D printed products with photopolymerization possess poor mechanical properties. An understanding of this is necessary for the advantages of 3D printing to be fully realized.

The mechanical properties of the 3D printed photopolymer were investigated with thermomechanical analysis and tensile testing. It has been found that the printed specimens are not fully cured after the 3D printing with photopolymerization. The DiBenedetto equation was used to better understand the relationship between the curing status and tensile properties (156).

Equations were derived from DiBenedetto (157) that express the effects of molecular weight, plasticization, degree of crosslinking, and copolymerization on the second order (i.e., glass) transition temperature. In their limits, the equations were shown to reduce in form to equations derivable from free volume theory. They have also been used to successfully analyze a variety of glass transition temperature data available in the literature on homogeneous un-crosslinked and crosslinked polymers, plasticized polymers, and random copolymers.

In addition to the poor mechanical properties, anisotropic and size-dependent tensile properties of the 3D printed photopolymers were also observed. An electron beam treatment can be used to ensure the curing of the 3D printed photopolymer (156).

2.22.2 *High-Performance Photopolymer with Low Volume Shrinkage*

A photocurable liquid low volume shrinkage photopolymer composition includes any type of polyfunctional (meth)acrylate monomer having two or more functionalities. The polyfunctional monomer serves to enhance the curing rate, adjust viscosity, and improve toughness of the 3D printed product.

A masterbatch composition is shown in Table 2.10.

Table 2.10 Masterbatch composition (158).

Compound	Amount/[%]		
Epoxy monomer	30	—	50
Polyfunctional (meth)acrylate	5	—	25
Space-filling monomer	30	—	50
Photoinitiator	0.5	—	2
Cationic initiator	0.5	—	2
Co-initiator	0.5	—	2
Light stabilizer	0.01	—	2

The masterbatch composition was made up by mixing the components well under subdued light. Specifically, a composition containing 40.84% pentaerythritol glycidyl ether, 14.90% trimethyolpropane trimethacrylate, 40.84% 2-phenoxyethylacrylate, 0.74% 2,4,6-trimethylbenzoyl diphenyl phosphine oxide (TPO), 0.88% bis(4-methylphenyl)iodonium hexafluorophosphate, 1.33% *N*-vinylcarbazole, and 0.47% Sudan I dye is prepared.

The photopolymer composition was cured using 3D printer. The characterized properties of the cured photopolymer are shown in Table 2.11.

Table 2.11 Properties (158).

Property	Value	Unit
Viscosity	70–80	cP
Hardness	78	D
Density	1.078	$g\,ml^{-1}$
Polymer density	1.132	$g\,ml^{-1}$
Volume shrinkage	4.78	%

Also, a composition containing 36.29% pentaerythritol tetramethacrylate, 30.92% stearyl methacrylate, 30.92% 2-hydroxybutyl acrylate, 1.84% 2,4,6-trimethylbenzoyl diphenyl phosphine oxide, cf. Figure 2.13, and 0.03% Sudan I dye was prepared. The photopolymer composition was cured using a 3D printer. The test results are shown in Table 2.12.

Table 2.12 Properties (158).

Property	Value	Unit
Viscosity	70–80	cP
Hardness	70	D
Density	1.091	$g\,ml^{-1}$
Polymer density	1.193	$g\,ml^{-1}$
Volume shrinkage	8.54	%
Ash content	0.098	%

2.22.3 *Dual Initiation Wavelengths for 3D Printing*

In the past, the focus of the SLA resins has been in the deep range at about 355 *nm*. These sources work well and there are many formulations for these sources. However, lasers at 355 *nm* are extremely expensive and rely on frequency tripled YVO4 laser crystal technology. Furthermore, digital-light-processing projectors are typically unreliable due to UV breakdown when farther away from visible

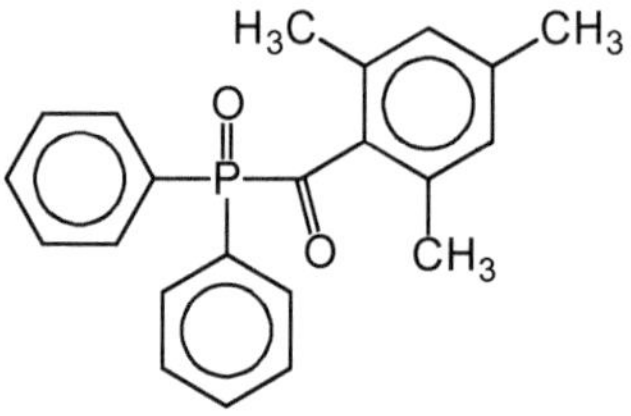

2,4,6-Trimethylbenzoyl diphenyl phosphine oxide

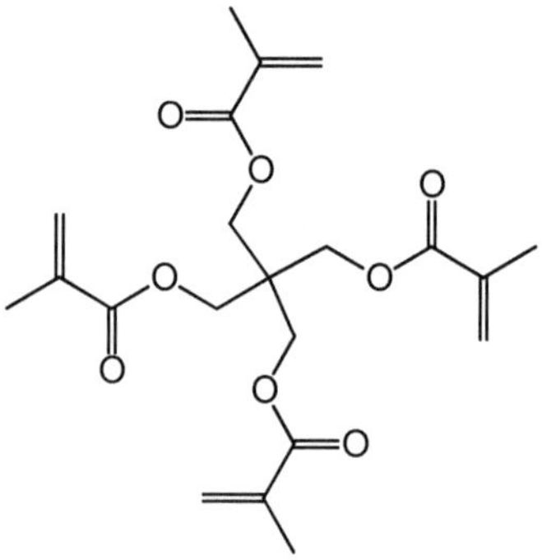

Pentaerythritol tetramethacrylate

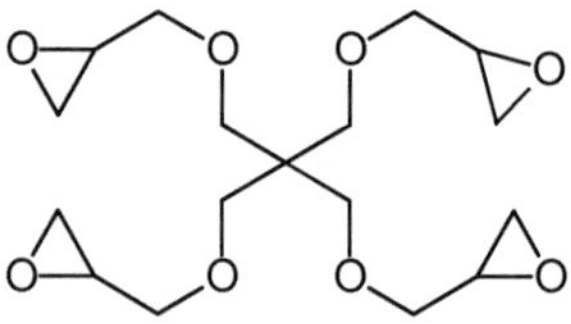

Pentaerythritol glycidyl ether

Figure 2.13 Photopolymer composition materials.

light and are not generally compatible with frequencies below 400 *nm*.

Due to the recent commercialization of Blu-ray laser diodes capable of directly emitting at 405 *nm* and production of 400 *nm* to 420 *nm* direct violet light-emitting diodes used in the production of white light bulbs, leading to low-cost light sources, there has been increased interest in creating SLA resins that can function with near-UV-sources in about the 400 *nm* to 420 *nm* range (159).

Additives comprising two photoinitiators with different absorption spectrums can be used to achieve printing and post-curing processes. A thermal initiator can also be used to complete the post-curing process by baking at appropriate temperature (160).

Additive combinations balanced to absorb some light in the near-UV range of about 400 *nm* to about 420 *nm* absorption wavelength band or spectrum while fluorescing light at frequencies higher than their absorption wavelength allow the creation of clear transparent materials using only near-UV light-sources (i.e., about 400 *nm* to about 420 *nm*) instead of deep UV (<400 *nm*) light sources (160).

An exemplary dual photoinitiator resin formulation has been described that contains about 68.50% by weight of Ebecryl 4858 (CAS: 120146-73-8) (a low viscosity aliphatic urethane diacrylate) (an oligomer), about 24.49% by weight of DPGDA (CAS: 57472-68-1) (dipropylene glycol diacrylate) (a monomer), about 4.95% by weight of Ebecryl 113 (CAS: 1204322-63-3) (a low odor monofunctional acrylated aliphatic epoxy) (a second monomer), about 0.98% by weight of Irgacure TPO (CAS: 75980-60-8) (acyl phosphine oxide photoinitiator), about 0.98% by weight of Irgacure 1173 (CAS: 7473-98-5) (2-Hydroxy-2-methyl-1-phenyl-propan-1-one) (a second photoinitiator) and about 0.10% weight of Benetex OB Plus (CAS: 7128-64-5) (a blocker) (2,2′-(2,5-thiophenediyl)bis(5-tert-butylbenzoxazole) (160).

Figure 2.14 illustrates an absorption spectrum of Irgacure TPO (CAS: 75980-60-8) and Irgacure 1173 (CAS: 7473-98-5).

2.23 Crosslinked Polymers

A polymer material for 3D printing has been described that has a first polymer and a second polymer (161). The first polymer and

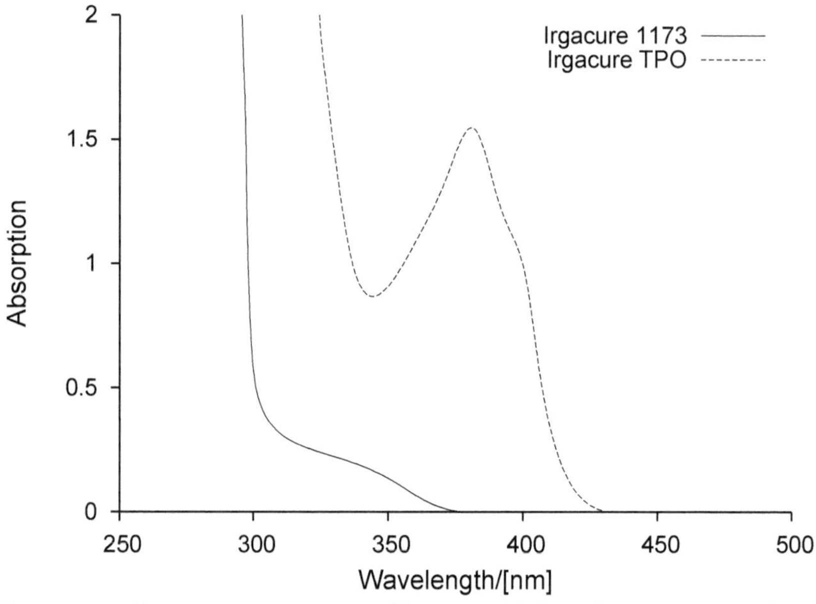

Figure 2.14 Absorption spectrum of Irgacure TPO and Irgacure 1173 (160).

the second polymer are crosslinked by a photo-crosslink, forming a polymer network. The first polymer and the second polymer are independently selected from PLA, ABS, PS, nylon, high density poly-(ethylene), PC, poly(vinyl alcohol), and poly(ethylene terephthalate) (PET).

Both polymers also include thermoplastic polymers such as styrenic block copolymers, thermoplastic olefins, elastomeric alloys, TPUs, thermoplastic copolyesters, and thermoplastic poly(amide)s (161).

Photocrosslinking can be done by irradiating the reaction mixture with ultraviolet light. Here, the reaction mixture must have one or more polymers with a photo-crosslinkable moiety. Suitable ultraviolet light sources include a light pipe, a fiber, and/or a bulb. A photo-crosslink may be photoreversible.

For example, a first polymer and second polymer that are crosslinked by a photo-crosslink may be irradiated with ultraviolet light to decompose the crosslink. The decomposition of a photo-crosslink yields a starting material such as a polymer having a photo-crosslinkable moiety. A photo-crosslink may be formed at a wavelength between about 200 *nm* and about 400 *nm*. The photo-crosslink unit may be photoreversible at a wavelength between this above-mentioned range. As photocrosslinking unit, an anthracene dimer, cinnamic acid dimer, coumarin dimer, thymine dimer, and stilbene dimer may be used.

The synthesis of a polymer with a crosslinkable moiety, i.e., a stilbene-substituted poly(lactide), may run as follows (161):

Preparation 2–1: Hydroxyl stilbene is treated with chloro alanine to yield an alanine-substituted stilbene. The hydrolysis of the amino moiety of the alanine-substituted stilbene may be accomplished by treating alanine-substituted stilbene with sodium nitrate and trifluoro acetic acid to yield the hydrolyzed-alanine-stilbene product. The hydrolyzed-alanine-stilbene product is then treated with 2-bromopropionyl bromide in the presence of triethylamine and 4-(dimethylamino)pyridine to result in a bromo-stilbene product. The bromo-stilbene product is treated with potassium iodide in the presence of acetone to yield an iodo-stilbene product. The iodo-stilbene product undergoes cyclization upon treatment with *N,N*-diisopropylethylamine in the presence of acetone to yield a stilbene-lactide product. The stilbene-lactide may undergo polymerization with lactide, for example, in the presence of dioctyl tin, to form a stilbene-substituted poly(lactide).

The compounds mentioned above are shown in Figure 2.15.

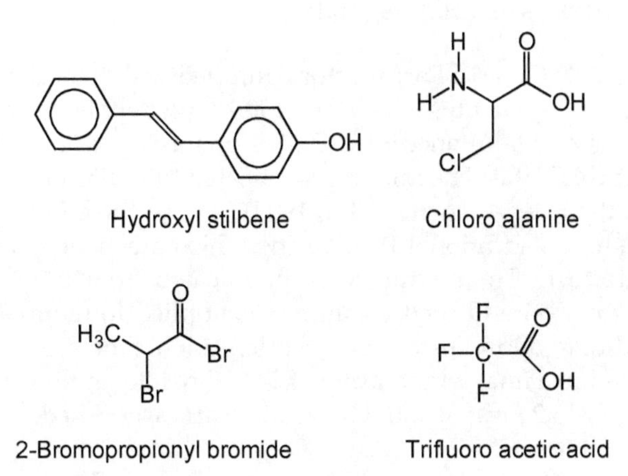

Figure 2.15 Chemicals of stilbene-substituted poly(lactide) synthesis.

2.24 Recycled Plastics

Copolymer resins for use in 3D printing technology have been described (162). These copolymers are low cost and can be primarily synthesized from oligomers obtained from recycled plastics and biobased monomers including diols, such as 1,4-butanediol, and diacids such as succinic acid.

The copolymers are accessible via depolymerizing PET plastic bottles, and varying amounts of diol and diacid. A particularly suitable starting material for the copolymers is a commercially available depolymerized product of recycled PET plastic bottles (Polylite, Reichhold Corp.), which is an oligomer with a molecular weight of about 800 *Dalton*.

Here, recycled PET can be reacted with succinic acid and 1,4-butanediol. The components are mixed in the presence of a tin-based catalyst FASCAT® 4100 at elevated temperatures to generate a copolymer. By varying the ratio of PET with 1,4-butanediol and succinic acid, numerous copolymers can be obtained. These materials each exhibit a wide array of physical characteristic features, while providing environmentally friendly products.

The resultant copolymers may be over 80% derived from sustainable monomers and provide better elongation at break and superior fatigue resistance. The copolymers are particularly suitable for 3D printing via fused deposition modeling processes. An example for the preparation is as follows (162):

Preparation 2–2: To a 1 *l* Parr reactor equipped with a mechanical stirrer and a distillation apparatus was added 150.27 *g* of depolymerized recycled PET, 274.94 *g* of 1,4-butanediol, 343.06 *g* of succinic acid and 2 *g* of tin catalyst FASCAT 4100. The mixture was heated to 160 °C under a nitrogen purge, and then slowly increased to 190 °C over a three hour period and maintained for an additional 18 *h*, during which time water was collected as the byproduct. The mixture was then heated from 190 °C to 200 °C over a 1.5 *h* period and then vacuum was applied to remove the excess 1,4-butanediol to allow further polycondensation. The mixture was then slowly heated to a final temperature of 240 °C, while under vacuum, until a viscosity of 313.2 *Poise* at 150 °C and 100 *rpm* was reached.

Then, resin filaments were made using a melt flow index instrument. A sample of the resin was melted in the heated barrel and extruded it through an orifice of a specific diameter under a certain weight. The above prepared sample showed a high flexibility with a breaking strain of 25.33%, which is close to the commercial PLA breaking strain of 26% (162).

2.25 3D Printed Fiber Reinforced Portland Cement Paste

Conventional Portland cement-based construction materials exhibit a high compressive strength of around 20 MPa to 60 MPa, but they fall short in terms of tensile and flexural strength values of 3 MPa to 10 MPa for plain cement pastes.

As a common solution, steel-reinforcement is placed in the formworks in order to improve flexural strength of the cementitious composite. However, steel-reinforcement results in time- and material-consuming labor costs during the construction process since the steel must be placed and fixed by hand in construction molds.

To avoid these disadvantages, mortars and concretes containing high-performance synthetic fibers, e.g., glass or carbon fibers, were developed (163). These composite materials showed a remarkable increase in the flexural and tensile properties, leading to an ultimate flexural strength of up to 50 MPa.

The properties of a 3D printed composite of Portland cement paste and reinforcing short fibers have been investigated (164). The properties of reinforcement fibers used are shown in Table 2.13.

Table 2.13 Properties of reinforcement fibers (164).

Fiber	Diameter [μm]	Tensile strength [MPa]	Young's modulus [GPa]	Length [mm]
Carbon fiber (HT C261)	7	3950	230	3
Glass fiber (AR Force D-6)	20	3500	72	6
Basalt fiber (BS 13 0064 12)	13	4200	93	6

The materials exhibit a high flexural strength of up to 30 MPa and a compressive strength of up to 80 MPa. An alignment of the fibers caused by the 3D printing process was observed, opening up the possibility of using the print path direction as a means to control fiber orientation within the printed structures. Apart from completely dense cementitious bodies, hierarchically structured bodies, displaying precisely adjusted macroporosity, have been developed. The latter materials exhibit a unique combination of strength and materials efficiency (164).

2.26 Polymer-Derived Ceramics

The issues with polymer-derived ceramics have historically been that the polymer-to-ceramic conversion occurs with gas release, typically leading to cracks or pores, which make the direct conversion of a preceramic part to a dense ceramic virtually unachievable unless its dimension is typically below a few hundred micrometers, as in the case of fibers, coatings, or foams (165). Preceramic polymers were proposed as precursors for the fabrication of mainly Si-based advanced ceramics, generally denoted as polymer-derived ceramics. The polymer-to-ceramic transformation process enabled significant technological breakthroughs in ceramic science and technology, such as the development of ceramic fibers, coatings, or ceramics stable at ultrahigh temperatures (up to 2000°C) with respect to decomposition, crystallization, phase separation, and creep. In recent years, several important advances have been achieved such as the discovery of a variety of functional properties associated with polymer-derived ceramics.

There are a couple of techniques that are currently used in commercial applications. One is the technique used by Robocasting Enterprises where a ceramic slurry is squeezed out of an applicator similar to the application of toothpaste to a toothbrush. The deposition pattern is controlled by a 3D CAD file to produce an initial green body which then must be heat treated at very high temperatures to densify to the final ceramic. Ceramics made with the robocasting technique include traditional ceramics such as alumina, zirconia, silicon nitride, and silicon carbide.

Another technology is one that is represented by 3DCeram where high viscosity, ultraviolet light-curable materials in paste form are used. These photocurable resin compounds containing ceramic powders are laid down in a manner such that a laser that is controlled by a 3D CAD file can polymerize the pastes. Then, another ceramic-UV curable paste layer is laid down on top of the previous layer followed by another laser treatment controlled by the 3D CAD file.

This process is repeated until the final 3D shape is obtained. The parts are then heat treated for the purpose of debinding the photocurable resin and then sintering the ceramic particles in order to eliminate the resin and densify the ceramic. Again, extreme temperatures in excess of 1600°C for long intervals are required to sinter the ceramics together. This high temperature process is very energy intensive, thus very expensive, and limits the composites that can be made with these techniques due to the temperatures required to sinter ceramics being generally higher than the melting temperature of most metals.

A recent advancement in additive manufacturing or 3D printing of ceramics or ceramic composites has been reported (166). Here, the fabrication of fully dense ceramic structures in intricate shapes have been detailed, such as rib, corkscrew, lattice and honeycomb, with no porosity or surface cracks, by using ultraviolet light-curable liquid polymer resins and exposing the liquid resin to UV light through a patterned mask using self-propagating photopolymer waveguide technology to rapidly create structures 100 to 1000 times more rapidly than with traditional layering. The architecture of the structure is defined by a patterned mask that defines the areas exposed to a collimated UV light source. To avoid shattering upon pyrolysis, the printed polymer structure is typically limited to fine features with less than approximately 3 *mm* thickness in one dimension. However, the size limitations of the structure are a drawback (167).

2.26.1 *Photocurable Ceramic/Polymer Composites*

Silane coupling agents with different organofunctional groups, cf., Table 2.14 and Figure 2.16, were coated on the surfaces of Al_2O_3 ceramic particles through hydrolysis and condensation reactions (168). The silane coupling agent-coated Al_2O_3 ceramic particles were dispersed in a commercial photopolymer-based on interpenetrating networks.

The organofunctional groups that have high radical reactivity and are more effective in UV curing systems are usually func-

Table 2.14 Silane coupling agents (168).

Short	Chemical name
VTMS	Vinyltrimethoxysilane
AMTMS	Acryloxymethyl trimethoxysilane
APTMS	3-Acryloxypropyl trimethoxysilane
MAPTMS	3-Methacryloxypropyl trimethoxysilane
ALPTMS	3-Acrylamidopropyl trimethoxysilane
AMPTMS	(Acryloxymethyl)phenethyl trimethoxysilane

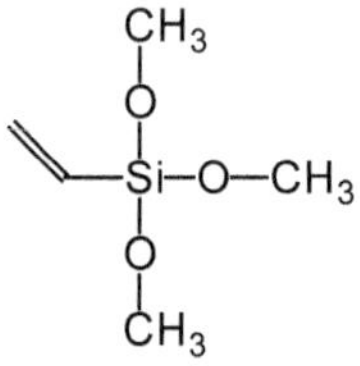

Vinyltrimethoxysilane

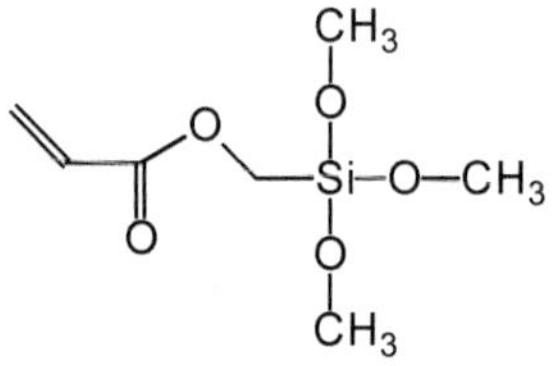

Acryloxymethyl trimethoxysilane

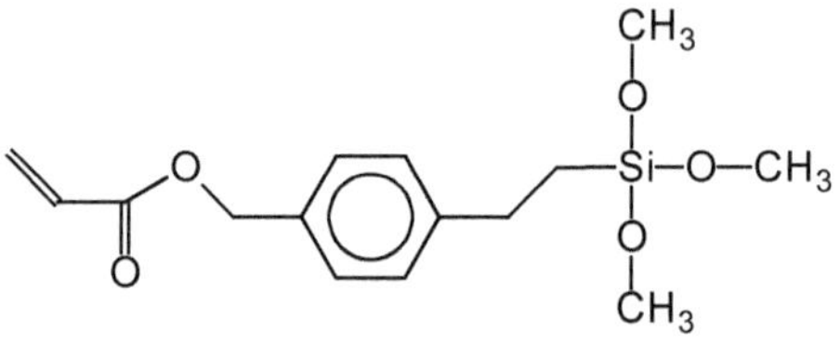

Acryloxymethyl)phenethyl trimethoxysilane

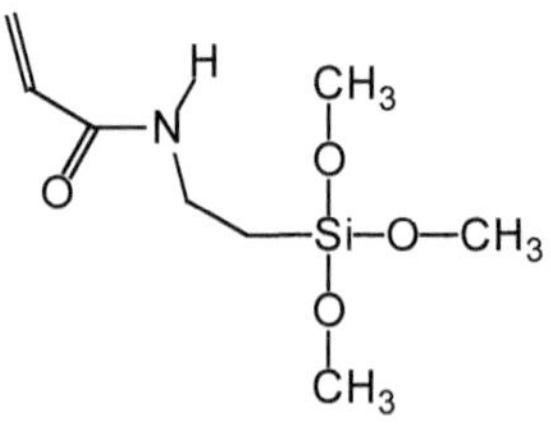

3-Acrylamidopropyl trimethoxysilane

Figure 2.16 Silane coupling agents.

tional groups based on acryl, such as acryloxy groups, methacrloxy groups, and acrylamide groups, and these silane coupling agents seem to improve interfacial adhesion and dispersion stability.

The coating morphology and the coating thickness distribution of silane coupling agent-coated Al_2O_3 ceramic particles according to the different organofunctional groups were observed by field-emission transmission electron microscopy. The initial dispersibility and dispersion stability of the silane coupling agent-coated Al_2O_3/high-temp composite solutions were investigated by relaxation nuclear magnetic resonance spectroscopy and Turbiscan. In addition, the rheological properties of the composite solutions were investigated by viscoelastic analysis and the mechanical properties of 3D printed objects were observed with a nanoindenter (168).

2.26.2 *Ceramic Matrix Composite Structures*

The particular attraction of preceramic polymers lies in the possibility of combining the properties of a polymeric feedstock very favorable for high-resolution additive buildup of parts with the capability of transforming them into a ceramic. Preceramic polymers are a special class of inorganic polymers that can convert with a high yield into ceramic materials, or polymer-derived ceramics, via high-temperature treatment in inert or oxidative atmospheres. The polymer-to-ceramic conversion occurs with gas release and shrinkage at 400°C–800°C (169).

The most frequently used preceramic polymers contain silicon atoms in the backbone, e.g., poly(siloxane)s, poly(silazane)s, and polycarbosilanes, yielding SiOC, SiCN, or SiC ceramics after pyrolysis. However, aluminum- and boron-containing polymers also are possible. In addition, preceramic polymers can be mixed with various fillers (either reactive or inert) to produce numerous advanced ceramic phases. This unique spectrum of characteristics has recently stimulated a variety of approaches for the use of preceramic polymers, either pure or mixed with fillers, as feedstocks in virtually all additive manufacturing technologies, both direct and indirect (170).

Preceramic polymers can allow fabrication of high-resolution, high-performance, and complex ceramic parts with an ease not encountered when processing powder-based systems.

An ink composition has been developed that contains chopped fibers that is suitable for direct ink writing (169,171). This composite enables the formation of ceramic matrix composite structures with complex shapes.

Both a preceramic polymer as polymeric binder and a ceramic source is used (171). Inks suitable for the extrusion of fine filaments with a diameter of smaller than 1 *mm* with a relatively high amount of fibers of more than 30% by volume for a nozzle diameter of 840 μm, have been formulated. Complex ceramic matrix composite structures with a porosity of around 75% and a compressive strength of around 4 *MPa* could be successfully printed.

Such a process is of particular interest because of its ability to orient the fibers in the extrusion direction due to the shear stresses generated at the nozzle tip. This phenomenon was observed in the production of polymer matrix composites, but has been employed for the first time for the production of ceramic matrices.

The possibility of aligning high aspect ratio fillers using direct ink writing opens a path to layer-by-layer design for optimizing the mechanical and microstructural properties within a printed object, and could potentially be extended to other types of fillers (171).

2.26.2.1 *Multiple Metals Doped Polymer-Derived Ceramics*

Multiple metals doped polymer-derived SiOC ceramics with an octet truss structure were prepared by employing a photosensitive methyl-silsesquioxane as preceramic polymer using a sol-gel method and digital light processing 3D printing (172).

The preparation of the composition runs as follows (172):

Preparation 2–3: A commercially available methyl-silsesquioxane resin, SILRES MK (Wacker Chemie GmbH, Germany) and MEMO, i.e., 3-(trimethoxysilyl) propyl methacrylate, were dissolved by adding in a mixture of tetrahydrofuran and tripropylene glycol monomethyl ether solvents in a 1:1 volume ratio. After 1 *h* of stirring, the mixture was hydrolyzed in acidic conditions at a pH of 2–3 at room temperature for 20 *min* to obtain a prehydrolyzed solution. Meanwhile, the mixture of zirconium *n*-propoxide and titanium isopropoxide at a 1:1 *M* ratio was stirred with methylacrylic acid as a complexing agent controlling the hydrolysis rate of metal alkoxide for 45 *min* to get the second solution. Then the first solution was added into the second solution slowly and continuously stirred for 5 *h*.

Before curing, trimethylol propane triacrylate was added as reactive diluent to reduce the viscosity and enhance the curing strength of the prepared resin. Then the tetrahydrofuran solvent was evaporated through the rotavapor for 60 *min* at 40°C and 6.324 *kPa* with a negative pressure of -0.095 *MPa* on the pressure header. As photoinitiator, 2,4,6-trimethylbenzoyl) phosphine oxide was added into the resin. The photoinitiator would experience a hemolytic cleavage and form radicals that could initiate the radical polymerization once exposed to light in the near-UV range. The curing dose and exposure time of the LED curing system could be adjusted according to the experimental requirements.

After 3D printing, the green body was carefully removed from the substrate and pyrolyzed at temperatures ranging from 500°C to 1200°C for 1 *h* in 99.99% nitrogen atmosphere with a heating rate of 2°C min^{-1}.

The influence of the photoinitiator on the photocuring property of the organosilicon resin is shown in Figure 2.17.

As the additive amount of the photoinitiator increases, the transmission depth coefficient decreases remarkably. The critical solidification energy has tended lower, however, it increases when the percentage of photoinitiator increases from 2% to 3%, which is thought to result from the nonuniform distribution of photoinitiator. The lowest critical exposure energy reaches about 5 $mJ\,cm^{-2}$, which demonstrates the excellent photocuring property of the synthesized resins.

Also, the reactive diluent trimethylol propane triacrylate significantly influences the viscosity of the resin. Before rotary evaporation, all the preceramic polymers have low viscosities from 50 to 80 *mPa s* and the influence of trimethylol propane triacrylate is not obvious. After rotary evaporation, the preceramic polymers still have good mobility because of the existence of trimethylol propane triacrylate, which is beneficial to the rapid leveling during the stereolithography 3D printing process. As the TMPTA amount increases from 5.71% to 17.14%, the viscosity of preceramic polymers reduces quickly to about 53%.

The physical and chemical properties of the preceramic polymers and printed octet truss structure SiOC ceramics were investigated. The results showed that the organosilicon preceramic polymers exhibit outstanding photocuring properties and could transform into amorphous SiOC ceramics at 800°C to 1200°C.

The excellent mechanical properties of SiOC ceramics with an octet truss structure, after 3D printing and pyrolysis, were attributed

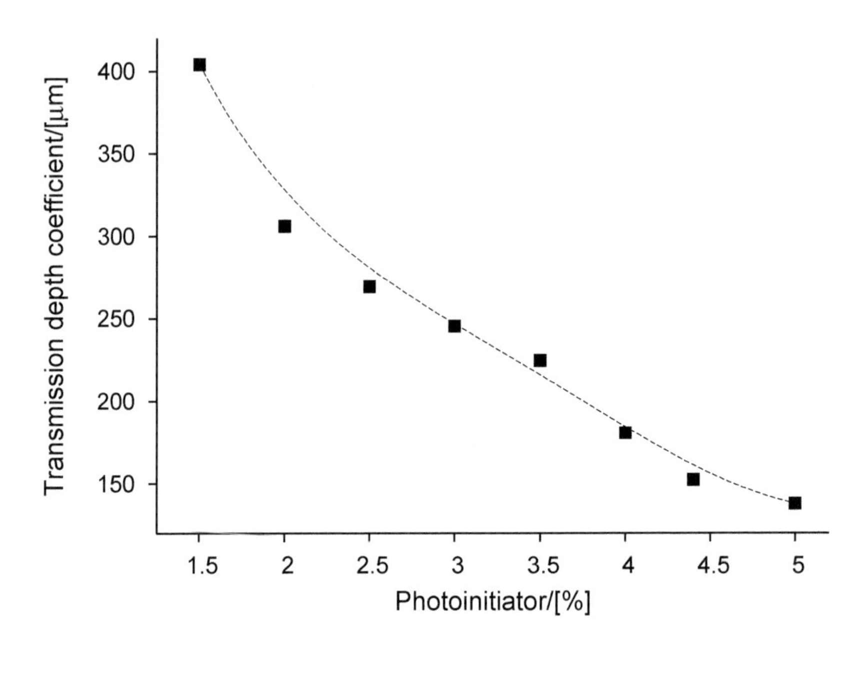

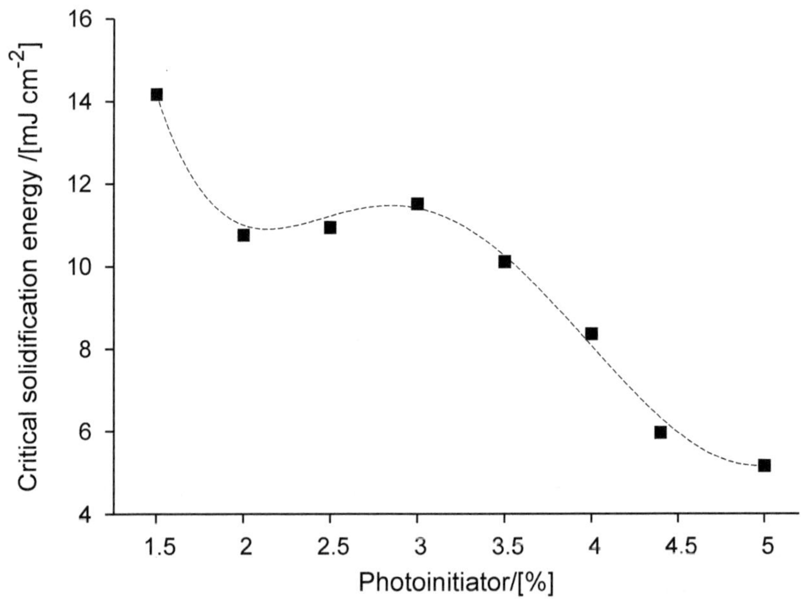

Figure 2.17 Influence of the photoinitiator on photocuring (172).

to the metal elements pinning in the amorphous matrix on the atomic level (172).

2.26.2.2 *SiCN Cellular Structures*

A simple method has been reported for generating complex and precise SiCN ceramic lattices using a preceramic polymer and applying the replica approach to structures fabricated using stereolithography of plastic materials, with the associated ease of fabrication (173).

The crosslinking of the preceramic polymer, poly(silazane), was promoted by adding a Pt-catalyst complex (platinum-divinyltetramethyldisiloxane) in xylene.

Three-dimensional printed plastic lattices impregnated with a poly(silazane) were converted to SiCN by pyrolysis at 1000°C in argon. In spite of the high amount of mass loss of around 58% and a volume shrinkage of 65%, the impregnated structures did not collapse during pyrolysis, leading to highly porous components with a total porosity ca. 93% per volume that possess suitable strength for handling and a potential for use as lightweight components (173).

2.26.2.3 *Monomers for Preceramic Polymers*

Resin formulations have been described, which may be used for 3D printing and pyrolyzing to produce a ceramic matrix composite (174). The resin formulations contain a solid-phase filler to provide high thermal stability and mechanical strength, e.g., fracture toughness, in the final ceramic material.

There are a wide variety of known preceramic polymers. Examples include poly(silazane)s, borazine-modified hydrido poly(silazane)s, polysilanes, polycarbosilanes, silicone resins, poly(vinylborazine), poly(borazylene), and decaborane-based polymers. These preceramic polymers have been used to form specific polymer-based structures that can be subsequently heat treated (pyrolyzed or sintered) to create near net-shape ceramic structures.

Some materials are shown in Table 2.15.

Also, a photoinitiator is added to the composition. Photoinitiators are shown in Table 2.16 and in Figures 2.18 and 2.19.

Table 2.15 Materials for preceramic polymers (174).

Compound
Trivinylborazine
2,4,6-Trimethyl-2,4,6-trivinylcyclotrisilazane
1,3,5,7-Tetravinyl-1,3,5,7-tetramethylcyclotetrasilazane
1,3,5-Trivinyl-1,3,5-trimethylcyclosiloxane
1,3,5,7-Tetravinyl-1,3,5,7-tetramethylcyclotetrasiloxane
2,2,4,4,6,6-Hexakisallyloxyl-triazatriphosphinine
Tetraallyloxysilane
Vinyl-terminated polydimethylsiloxane
Tetravinylsilane
Vinyl-terminated dimethylsiloxane-ethylene copolymer
Divinyldimethylsilane
1,2-Divinyltetramethyldisilane
1,4-Bis(vinyldimethylsilyl)benzene
Vinylmethylsiloxane homopolymer
Methacryloxypropyl-terminated polydimethylsiloxane
Boron vinyldimethylsiloxide
Vinylmethylsiloxane-dimethylsiloxane copolymer
Trimethylsiloxy-terminated homopolymer
Vinylethoxysiloxane-propylethoxysiloxane copolymer
Vinyltrimethoxysilane
Trivinylmethylsilane
Diallyldimethylsilane
1,3,5-Trisilacyclohexane
B,B′B″-Trithynyl-N,N′N″-trimethylborazine
B,B′B″-Triethynylborazine
Vinylmethoxysiloxane
Acryloxypropyl methylsiloxane homopolymer

Table 2.16 Initiators for preceramic polymers (174).

Photoinitiator
2,2-Dimethoxy-2-phenyl-acetophenone
2-Hydroxy-2-methylpropiophenone
Camphorquinone
Bis(2,4,6-trimethylbenzoyl)-phenylphosphine oxide
Benzophenone
Dibenzoyl peroxide

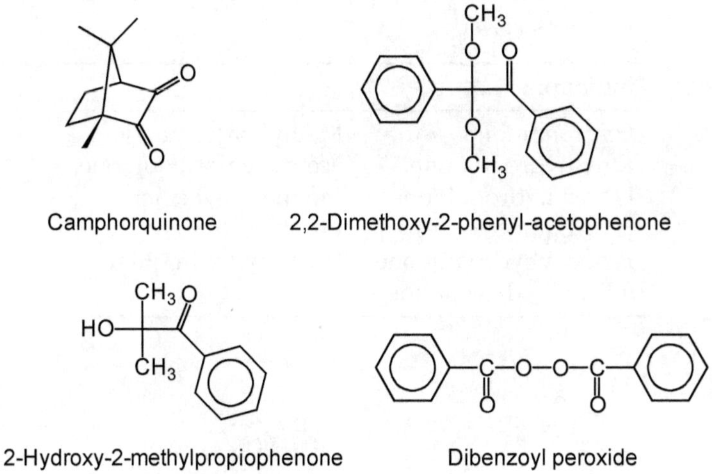

Camphorquinone 2,2-Dimethoxy-2-phenyl-acetophenone

2-Hydroxy-2-methylpropiophenone Dibenzoyl peroxide

Figure 2.18 Photoinitiators.

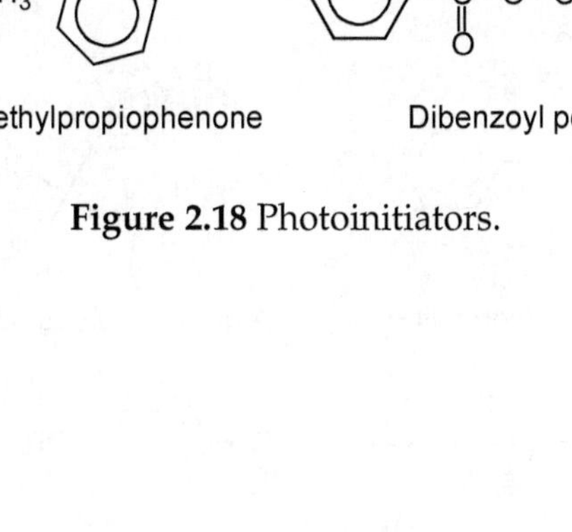

Figure 2.19 Bis(2,4,6-trimethylbenzoyl)-phenylphosphine oxide.

Also, a free-radical inhibitor may be present in a concentration from about 0.001% to about 10% in the formulation. Free-radical inhibitors are shown in Table 2.17 and in Figure 2.20.

Table 2.17 Free-radical inhibitors for preceramic polymers (174).

Inhibitor	
Hydroquinone	Methylhydroquinone
Ethylhydroquinone	Methoxyhydroquinone
Ethoxyhydroquinone	Monomethyl ether
Hydroquinone	Propylhydroquinone
Propoxyhydroquinone	*tert*-Butylhydroquinone
n-Butylhydroquinone	

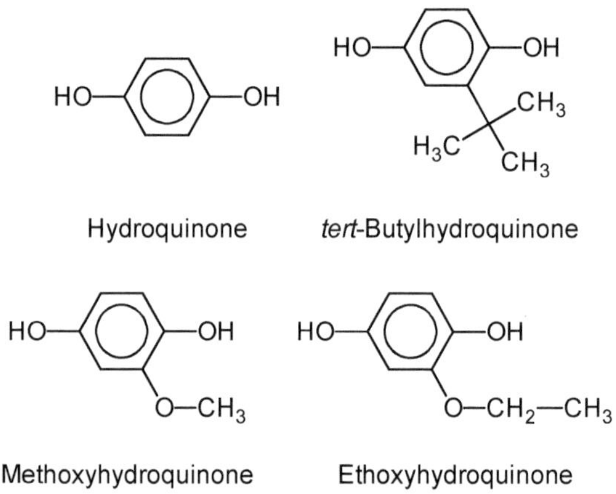

Hydroquinone *tert*-Butylhydroquinone

Methoxyhydroquinone Ethoxyhydroquinone

Figure 2.20 Free-radical inhibitors.

A 3D printing resolution agent may be present in a concentration from about 0.001% to about 10% in the formulation. 3D printing resolution agents are shown in Table 2.18.

To increase the fracture toughness of a 3D printed ceramic matrix composite, a high-aspect-ratio filler, such as a fiber, may be coated with a filler/matrix interfacial coating. The purpose of this coating is to provide a weak filler-matrix interface that prevents matrix cracks from penetrating the fillers, thus providing damage tolerance (toughness) to the composite. The interfacial coating is preferably

Table 2.18 3D printing resolution agents for preceramic polymers (174).

Resolution agent
2-(2-Hydroxyphenyl)-benzotriazole
2-Hydroxyphenyl-benzophenone
2-Hydroxyphenyl-*s*-triazines
2,2′-(2,5-Thiophenediyl)bis(5-tert-butylbenzoxazole)
2,2′-(1,2-Ethenediyl)bis(4,1-phenylene)bisbenzoxazole

chemically and mechanically stable during processing and pyrolysis. Examples of interfacial coatings include BN, C, AlN, or a combination thereof (174).

2.26.3 *Selective Laser Melting*

A process for forming a finished green body component in an additive manufacturing system has been described (167). It consists of the following steps:

1. Selecting a precursor resin,
2. Converting the precursor resin to beads,
3. Blending the precursor resin beads with a powder selected from at least one of a metal powder, a carbide powder, a ceramic powder and a mixture thereof,
4. Depositing a plurality of layers of the polymer precursor resin and powder blend in a bed,
5. Spraying each layer with photocurable or thermally curable resins,
6. Heating the layers and the entire bead bed with ultraviolet or infrared radiation to cure the resin mixture, and
7. Form a finished green body component,
8. Removing the finished green body component to a furnace to convert the green body to a ceramic composite having a thickness in a depth dimension in a range between approximately 200 μm and approximately 25 mm.

The step of depositing of the plurality of layers of the polymer precursor resin and powder blend can be computer controlled. The precursor resin can be selected from one of a liquid resin and a multiple of different precursor resins. The precursor resin can be enhanced with a plurality of enhancement particles selected from the group consisting of a metallic powder, ceramic powder, graphite powder, graphene powder, diamond powder, carbide powder, silicide powder, nitride powder, oxide powder, graphene, carbon nanofiber, carbon nanotubes, and mixtures thereof.

The objects can be of almost any shape or geometry that can be accommodated by the 3D printing process. The three-dimensional ceramic structure is provided by putting the finished green body component in a furnace to convert it to a ceramic piece, the resulting ceramic structure is a solid, monolithic piece having a minimum thickness of approximately 200 μm. If the object has a cube shape, the dimensions are approximately 200 μm in height, approximately 200 μm in width and approximately 200 μm in depth. If the shape is a three-dimensional panel, the maximum thickness is approximately 25 mm, the maximum height is approximately 1000 mm and the maximum width is approximately 1000 mm.

When a selective laser melting process is used to form the solid structures containing spherical ceramic beads in melted metallic or glass pastes, the size of the monolithic, solid structure produced is only limited by the size of the 3D printer. For example, a solid cube could have the dimensions of approximately 1000 $mm \times$ 1000 $mm \times$ 1000 mm. This is possible because there is no need for a furnace (167).

2.26.4 *Stereolithography Resin for Rapid Prototyping of Ceramics and Metals*

Photocurable ceramic resins having solid loadings in excess of 40% v/v and a viscosity of less than 3000 $mPa\,s$ are suitable for multilayer fabrication of green ceramic parts by stereolithography and similar techniques (175).

The green ceramic parts, which may be of traditional ceramic materials, sinterable metals, or combinations thereof, exhibit a low shrinkage upon firing or sintering, and may be used for such applications as rapid prototyping, biocompatible ceramic parts, ceramic cores for investment casting, ceramic molds for metal casting, and others (175).

References

1. T.I. Zohdi, Introduction: Additive/3D printing materials—filaments, functionalized inks, and powders in *Modeling and Simulation of Functionalized Materials for Additive Manufacturing and 3D Printing: Continuous and Discrete Media: Continuum and Discrete Element Methods*, chapter 1, pp. 1–7. Springer International Publishing, Cham, 2018.
2. X. Wang, M. Jiang, Z. Zhou, J. Gou, and D. Hui, *Composites Part B: Engineering*, Vol. 110, p. 442 , 2017.
3. Q. Sun, G.M. Rizvi, C.T. Bellehumeur, and P. Gu, *Rapid Prototyping Journal*, Vol. 14, p. 72, 2008.
4. B.M. Tymrak, M. Kreiger, and J.M. Pearce, *Materials & Design*, Vol. 58, p. 242, 2014.
5. P. Tran, T.D. Ngo, A. Ghazlan, and D. Hui, *Composites Part B: Engineering*, Vol. 108, p. 210, 2017.
6. R. Melnikova, A. Ehrmann, and K. Finsterbusch, *IOP Conference Series: Materials Science and Engineering*, Vol. 62, p. 012018, 2014.
7. B. Caulfield, P. McHugh, and S. Lohfeld, *Journal of Materials Processing Technology*, Vol. 182, p. 477, 2007.
8. C.R. Garcia, J. Correa, D. Espalin, J.H. Barton, R.C. Rumpf, R. Wicker, and V. Gonzalez, *Progress In Electromagnetics Research*, Vol. 34, p. 75, 2012.
9. J. Dou, Q. Zhang, M. Ma, and J. Gu, *Journal of Magnetism and Magnetic Materials*, Vol. 324, p. 3078, 2012.
10. H. Gu, C. Ma, J. Gu, J. Guo, X. Yan, J. Huang, Q. Zhang, and Z. Guo, *J. Mater. Chem. C*, Vol. 4, p. 5890, 2016.
11. J. Gu, X. Yang, Z. Lv, N. Li, C. Liang, and Q. Zhang, *International Journal of Heat and Mass Transfer*, Vol. 92, p. 15, 2016.
12. E. Kroll and D. Artzi, *Rapid Prototyping Journal*, Vol. 17, p. 393, 2011.
13. K.V. Wong and A. Hernandez, *ISRN Mechanical Engineering*, Vol. 2012, p. 1, 2012.
14. D.B. Short, *3D Printing and Additive Manufacturing*, Vol. 2, p. 209, 2015.
15. S.V. Murphy and A. Atala, *Nature Biotechnology*, Vol. 32, p. 773, August 2014.
16. N.S. Hmeidat, J.W. Kemp, and B.G. Compton, *Composites Science and Technology*, Vol. 160, p. 9, 2018.
17. S. Malhotra, K. Goda, and M.S. Sreekala, Introduction to polymer composites in S. Thomas, K. Joseph, S.K. Malhotra, K. Goda, and M.S. Sreekala, eds., *Polymer Composites*, chapter 1. Wiley-VCH, Weinheim, 1st edition, 2012.
18. S.H. Huang, P. Liu, A. Mokasdar, and L. Hou, *The International Journal of Advanced Manufacturing Technology*, Vol. 67, p. 1191, 2013.

19. C.A. Murphy and M.N. Collins, *Polymer Composites*, Vol. 39, p. 1311, 2018.
20. G. Postiglione, G. Natale, G. Griffini, M. Levi, and S. Turri, *Polymer Composites*, Vol. 38, p. 1662, 2015.
21. J. Wang, A. Chiappone, I. Roppolo, F. Shao, E. Fantino, M. Lorusso, D. Rentsch, K. Dietliker, C.F. Pirri, and H. Grützmacher, *Angewandte Chemie, International Edition*, Vol. 57, p. 2353, 2017.
22. Y. Han, F. Wang, H. Wang, X. Jiao, and D. Chen, *Composites Science and Technology*, Vol. 154, p. 104 , 2018.
23. Y. Zhao, W.N. Martens, T.E. Bostrom, H.Y. Zhu, and R.L. Frost, *Langmuir*, Vol. 23, p. 2110, 2007.
24. Wikipedia contributors, Hydrothermal synthesis — Wikipedia, the free encyclopedia, https://en.wikipedia.org/w/index.php?title=Hydrothermal_synthesis&oldid=788720494, 2017. [Online; accessed 17-May-2018].
25. D.E. Yunus, W. Shi, S. Sohrabi, and Y. Liu, *Nanotechnology*, Vol. 27, p. 495302, 2016.
26. L. Tse, S. Kapila, and K. Barton, Contoured 3D printing of fiber reinforced polymers, in *Solid Freeform Fabrication 2016: Proceedings of the 27th Annual International Solid Freeform Fabrication Symposium*, pp. 1205–1216. University of Texas, 2016.
27. J.P. Lewicki, J.N. Rodriguez, C. Zhu, M.A. Worsley, A.S. Wu, Y. Kanarska, J.D. Horn, E.B. Duoss, J.M. Ortega, and W. Elmer, *Scientific Reports*, Vol. 7, p. 43401, 2017.
28. K.C. Lin and C.M. Hsieh, *Composite Structures*, Vol. 79, p. 606, 2007.
29. W. Hao, D. Ge, Y. Ma, X. Yao, and Y. Shi, *Polymer Testing*, Vol. 31, p. 520, 2012.
30. W. Hao, Y. Liu, H. Zhou, H. Chen, and D. Fang, *Polymer Testing*, Vol. 65, p. 29, 2018.
31. A.K. Sood, R.K. Ohdar, and S.S. Mahapatra, *Materials & Design*, Vol. 31, p. 287, 2010.
32. Stratasys, Vantage user guide, version 1.1. 2004, electronic: http://www.stratasys.com/, 2004.
33. D. Yang, K. Wu, L. Wan, and Y. Sheng, *Journal of Manufacturing and Materials Processing*, Vol. 1, 2017.
34. K. Wu, D. Yang, and N. Wright, *Computers & Structures*, Vol. 177, p. 141, 2016.
35. K. Wu, D. Yang, N. Wright, and A. Khan, *Journal of Fluids and Structures*, Vol. 76, p. 166, 2018.
36. S. Hwang, E.I. Reyes, K.-S. Moon, R.C. Rumpf, and N.S. Kim, *Journal of Electronic Materials*, Vol. 44, p. 771, Mar 2015.

37. E.M. Sachs, J.S. Haggerty, M.J. Cima, and P.A. Williams, Three-dimensional printing techniques, US Patent 5 204 055, assigned to Massachusetts Institute of Technology (Cambridge, MA), April 20, 1993.
38. B. Utela, D. Storti, R. Anderson, and M. Ganter, *Journal of Manufacturing Processes*, Vol. 10, p. 96, 2008.
39. F.P.W. Melchels, J. Feijen, and D.W. Grijpma, *Biomaterials*, Vol. 31, p. 6121, 2010.
40. Y.H. Cho, I.H. Lee, and D.-W. Cho, *Microsystem Technologies*, Vol. 11, p. 158, February 2005.
41. C. Heller, M. Schwentenwein, G. Russmueller, F. Varga, J. Stampfl, and R. Liska, *Journal of Polymer Science Part A: Polymer Chemistry*, Vol. 47, p. 6941, 2009.
42. D.D. Gu, W. Meiners, K. Wissenbach, and R. Poprawe, *International Materials Reviews*, Vol. 57, p. 133, 2012.
43. I. Gibson and D. Shi, *Rapid Prototyping Journal*, Vol. 3, p. 129, 1997.
44. G. Griffini, M. Invernizzi, M. Levi, G. Natale, G. Postiglione, and S. Turri, *Polymer*, Vol. 91, p. 174 , 2016.
45. U. Kalsoom, A. Peristyy, P.N. Nesterenko, and B. Paull, *RSC Adv.*, Vol. 6, p. 38140, 2016.
46. P. Tummala, P.S. Turner, and M.A. Johnson, Adhesive for 3D printing, US Patent 9 757 881, assigned to 3D Systems, Inc. (Rock Hill, SC), September 12, 2017.
47. Air Products, Air products' specialty additives. to improve your waterborne wood coatings quality with ease, electronic: http://www.airproducts.com/~/media/Files/PDF/microsites/wood-coating/en-PMD-SPA-Wood-Coating-updated.pdf?la=en, 2014.
48. D. Lin, S. Jin, F. Zhang, C. Wang, Y. Wang, C. Zhou, and G.J. Cheng, *Nanotechnology*, Vol. 26, p. 434003, 2015.
49. Cheaptubes, Single-layer graphene oxide, electronic: https://www.cheaptubes.com/product/single-layer-graphene-oxide/, 2018.
50. EnvisionTec, PIC 100 series, electronic: https://envisiontec.com/3d-printing-materials/micro-materials/pic100series/, 2017.
51. S. Yuan, Y. Zheng, C.K. Chua, Q. Yan, and K. Zhou, *Composites Part A: Applied Science and Manufacturing*, Vol. 105, p. 203, 2018.
52. EOS, Product information PA 2200 – polyamide white, electronic: https://www.eos.info/pa-2200-2201-da42d3327dfc8bf2, 2018.
53. Bayer MaterialScience, New thermoplastic polyurethanes for laser sintering, electronic: http://www.prototypetoday.com/news/bayer-materialscience, 2018.
54. J.K. Hohmann, M. Renner, E.H. Waller, and G. von Freymann, *Advanced Optical Materials*, Vol. 3, p. 1488, 2015.

55. U. Kalsoom, P.N. Nesterenko, and B. Paull, *RSC Adv.*, Vol. 6, p. 60355, 2016.
56. R.D. Farahani, M. Dubé, and D. Therriault, *Advanced Materials*, Vol. 28, p. 5794, 2016.
57. R.D. Farahani and D. Martine, *Advanced Engineering Materials*, Vol. 20, p. 1700539, 2017.
58. W.E. Jones, J. Chiguma, E. Johnson, A. Pachamuthu, and D. Santos, *Materials*, Vol. 3, p. 1478, February 2010.
59. C.S. Boland, U. Khan, G. Ryan, S. Barwich, R. Charifou, A. Harvey, C. Backes, Z. Li, M.S. Ferreira, M.E. Möbius, R.J. Young, and J.N. Coleman, *Science*, Vol. 354, p. 1257, 2016.
60. S.J. Leigh, C.P. Purssell, J. Bowen, D.A. Hutchins, J.A. Covington, and D.R. Billson, *Sensors and Actuators A: Physical*, Vol. 168, p. 66, 2011.
61. S.C. Tjong, *Materials Science and Engineering: R: Reports*, Vol. 53, p. 73, 2006.
62. L.L. Lebel, B. Aissa, M.A.E. Khakani, and D. Therriault, *Composites Science and Technology*, Vol. 70, p. 518, 2010.
63. R.D. Farahani, H. Dalir, V.L. Borgne, L.A. Gautier, M.A.E. Khakani, M. Lévesque, and D. Therriault, *Composites Science and Technology*, Vol. 72, p. 1387, 2012.
64. E.T. Thostenson and T.-W. Chou, *Carbon*, Vol. 44, p. 3022, 2006.
65. S. Abbasi, P.J. Carreau, and A. Derdouri, *Polymer*, Vol. 51, p. 922, 2010.
66. A.I. Alateyah, H.N. Dhakal, and Z.Y. Zhang, *Advances in Polymer Technology*, Vol. 32, p. 21368, 2013.
67. A.M. Díez-Pascual, M. Naffakh, C. Marco, G. Ellis, and M.A. Gómez-Fatou, *Progress in Materials Science*, Vol. 57, p. 1106, 2012.
68. C. Al de Leon, Q. Chen, N.B. Palaganas, J.O. Palaganas, J. Manapat, and R.C. Advincula, *Reactive and Functional Polymers*, Vol. 103, p. 141, 2016.
69. J. Hiemenz, Additive manufacturing trends in aerospace, http://3dsolids.co.za/PDF/White%20Papers/AeroTrends.pdf, 2014.
70. K.C. Chuang, J.E. Grady, R.D. Draper, E.-S.E. Shin, C. Patterson, and T.D. Santelle, Additive manufacturing and characterization of Ultem polymers and composites, in *The Composites and Advanced Materials Expo CAMX*, Dallas, TX, 2015. Society for the Advancement of Materials and Process Engineering, NASA Glenn Research Center, Cleveland, OH.
71. Y. Huang, M.C. Leu, J. Mazumder, and A. Donmez, *Journal of Manufacturing Science and Engineering*, Vol. 137, p. 014001, February 2015.
72. T.D. Ngo, A. Kashani, G. Imbalzano, K.T. Nguyen, and D. Hui, *Composites Part B: Engineering*, Vol. 143, p. 172, 2018.

73. S.C. Ligon, R. Liska, J. Stampfl, M. Gurr, and R. Mülhaupt, *Chemical Reviews*, Vol. 117, p. 10212, 2017.
74. V. Kunc, O. Rios, L.J. Love, C.E. Duty, and A. Johs, Reactive polymer fused deposition manufacturing, US Patent 9 650 537, assigned to UT-Battelle, LLC (Oak Ridge, TN), May 16, 2017.
75. C.W. Hull, Apparatus for production of three-dimensional objects by stereolithography, US Patent 4 575 330, assigned to UVP, Inc. (San Gabriel, CA), March 11, 1986.
76. S.K. Mirle and R.J. Kumpfmiller, Photosensitive compositions useful in three-dimensional part-building and having improved photospeed, US Patent 5 418 112, assigned to W. R. Grace & Co.-Conn. (New York, NY), May 23, 1995.
77. P. Jacobs, *Rapid Prototyping & Manufacturing: Fundamentals of Stereolithography*, Society of Manufacturing Engineers in cooperation with the Computer and Automated Systems Association of SME, Dearborn, MI, 1992.
78. W. Green, *Industrial Photoinitiators: A Technical Guide*, CRC Press, Boca Raton, 2010.
79. D.G. Leppard, M. Kohler, and L. Misev, Photopolymerizable compositions containing an alkylbisacylphosphine oxide, US Patent 5 472 992, assigned to Ciba-Geigy Corporation (Ardsley, NY), December 5, 1995.
80. E. Fodran, M. Koch, and U. Menon, Mechanical and dimensional characteristics of fused deposition modeling build styles, in *Solid Freeform Fabrication Proceedings*, pp. 419–442, 1996.
81. G.L. Collins and J.R. Costanza, *Journal of Coatings Technology*, Vol. 51, p. 57, 1979.
82. D.D.M. Wayner, K.B. Clark, A. Rauk, D. Yu, and D.A. Armstrong, *Journal of the American Chemical Society*, Vol. 119, p. 8925, 1997.
83. H. Bayer, W. Fischer, V. Muhrer, W. Rogler, and L. Schön, Low-shrinkage light-curable resin, WO Patent 1 997 016 482, assigned to Siemens Aktiengesellschaft, May 09, 1997.
84. B. Steinmann, Stereolithographic resins with high temperature and high impact resistance, US Patent 6 989 225, assigned to 3D Systems, Inc. (Valencia, CA), January 24, 2006.
85. T.Y. Lee, C.A. Guymon, E.S. Jönsson, and C.E. Hoyle, *Polymer*, Vol. 45, p. 6155, 2004.
86. W.D. Cook, *Journal of Polymer Science Part A: Polymer Chemistry*, Vol. 31, p. 1053, 1993.
87. E.J. Murphy, R.E. Ansel, and J.J. Krajewski, Method of forming a three-dimensional object by stereolithography and composition therefore, US Patent 4 942 001, assigned to DeSoto and Inc. (Des Plaines, IL), July 17, 1990.

88. A.L. Coats, J.P. Harrison, J.S. Hay, and M.J. Ramos, Stereolithography resins and methods, US Patent 7 211 368, assigned to 3 Birds, Inc. (Pompano Beach, FL), May 1, 2007.
89. B. Modrek, B. Parker, and S.T. Spence, Methods for curing partially polymerized parts, US Patent 5 164 128, assigned to 3D Systems, Inc. (Valencia, CA), November 17, 1992.
90. B. Steinmann, J.-P. Wolf, A. Schulthess, and M. Hunziker, (Meth)-acrylates containing urethane groups, US Patent 5 658 712, assigned to Ciba-Geigy Corporation (Tarrytown, NY), August 19, 1997.
91. B. Steinmann and A. Schulthess, Liquid, radiation-curable composition, especially for stereolithography, US Patent 5 972 563, assigned to Ciba Specialty Chemicals Corp. (Tarrytown, NY), October 26, 1999.
92. M. Ueda, K. Takase, and T. Kurosawa, Stereolithography resin compositions and three-dimensional objects made therefrom, US Patent 8 980 971, assigned to DSM IP Assets B.V. (Heerlen, NL), March 17, 2015.
93. B. Steinmann, J.-P. Wolf, A. Schulthess, and M. Hunziker, Photosensitive compositions, US Patent 5 476 748, assigned to Ciba-Geigy Corporation (Tarrytown, NY), December 19, 1995.
94. L.U. Kim, J.W. Kim, and C.K. Kim, *Biomacromolecules*, Vol. 7, p. 2680, 2006.
95. C.W. Hull, S.T. Spence, C.W. Lewis, W. Vinson, R.S. Freed, and D.R. Smalley, Stereolithographic curl reduction, US Patent 5 772 947, June 30, 1998.
96. R.R. Moraes, J.W. Garcia, M.D. Barros, S.H. Lewis, C.S. Pfeifer, J. Liu, and J.W. Stansbury, *Dental Materials*, Vol. 27, p. 509, 2011.
97. B. Husár, S.C. Ligon, H. Wutzel, H. Hoffmann, and R. Liska, *Progress in Organic Coatings*, Vol. 77, p. 1789, 2014.
98. A.F. Jacobine, M.A. Rakas, and D.M. Glaser, Stereolithography method, US Patent 5 167 882, assigned to Loctite Corporation (Newington, CT), December 1, 1992.
99. C.E. Hoyle and C.N. Bowman, *Angewandte Chemie International Edition*, Vol. 49, p. 1540.
100. S.C. Ligon, B. Husár, H. Wutzel, R. Holman, and R. Liska, *Chemical Reviews*, Vol. 114, p. 557, 2014.
101. P. Esfandiari, S.C. Ligon, J.J. Lagref, R. Frantz, Z. Cherkaoui, and R. Liska, *Journal of Polymer Science Part A: Polymer Chemistry*, Vol. 51, p. 4261, 2013.
102. M. Edler, F.H. Mostegel, M. Roth, A. Oesterreicher, S. Kappaun, and T. Griesser, *Journal of Applied Polymer Science*, Vol. 134, p. 4493, 2017.
103. A. Oesterreicher, A. Moser, M. Edler, H. Griesser, S. Schlögl, M. Pichelmayer, and T. Griesser, *Macromolecular Materials and Engineering*, Vol. 302, p. 1600450, 2017.

104. J.R. Tumbleston, D. Shirvanyants, N. Ermoshkin, R. Janusziewicz, A.R. Johnson, D. Kelly, K. Chen, R. Pinschmidt, J.P. Rolland, A. Ermoshkin, E.T. Samulski, and J.M. DeSimone, *Science*, Vol. 347, p. 1349, 2015.
105. D.G. Sycks, T. Wu, H.S. Park, and K. Gall, *Journal of Applied Polymer Science*, Vol. 135, p. 46259, 2018.
106. D.G. Sycks, D.L. Safranski, N.B. Reddy, E. Sun, and K. Gall, *Macromolecules*, Vol. 50, p. 4281, 2017.
107. S.C. Lapin, J.R. Snyder, E.V. Sitzmann, D.K. Barnes, and G.D. Green, Stereolithography using vinyl ether-epoxide polymers, US Patent 5 437 964, assigned to AlliedSignal Inc. (Morris Township, Morris County, NJ), August 1, 1995.
108. J.V. Crivello and U. Varlemann, Structure and reactivity relationships in the photoinitiated cationic polymerization of 3,4-epoxycyclohexylmethyl-3′,4′-epoxycyclohexane carboxylate, in A.B. Scranton, C.N. Bowman, and R.W. Pheiffer, eds., *Photopolymerization: Fundamentals and Applications*, pp. 82–94, Washington, DC, 1997. 211th National Meeting of the American Chemical Society, New Orleans, American Chemical Society.
109. J.-P. Allen, P. Blondel, and P. Douais, Increase in the melting point and the enthalpy of melting of polyamides by a water treatment, US Patent 7 468 405, assigned to Atofina (Puteaux, FR), December 23, 2008.
110. H.K. Park, M. Shin, B. Kim, J.W. Park, and H. Lee, *NPG Asia Materials*, p. 1, 2018.
111. M. Rinaldi, T. Ghidini, F. Cecchini, A. Brandao, and F. Nanni, *Composites Part B: Engineering*, Vol. 145, p. 162, 2018.
112. ASTM International, Standard test method for tensile properties of plastics, ASTM Standard ASTM D638-14, ASTM International, West Conshohocken, PA, 2014.
113. C. Chua, *Lasers in 3D Printing and Manufacturing*, Vol. 2 of *World Scientific Series in 3D Printing*, World Scientific, New Jersey, 2017.
114. C. Song, A. Huang, Y. Yang, Z. Xiao, and J.-K. Yu, *Rapid Prototyping Journal*, Vol. 23, p. 1069, 2017.
115. J. Schmidt, M. Sachs, S. Fanselow, K.-E. Wirth, and W. Peukert, New approaches towards production of polymer powders for selective laser beam melting of polymers, in *AIP Conference Proceedings*, Vol. 1914, p. 190008. AIP Publishing, 2017.
116. D.S.-R. Liu, J.J. Reilly, and M.A. Aubart, Dimensionally stable acrylic alloy for 3-D printing, WO Patent 2 017 210 286, assigned to Arkema France, December 07, 2017.
117. A. Shulga, I. Kovalev, and D. Starodubtsev, Composition for 3D printing, US Patent 9 527 992, assigned to CreoPop Pte. Ltd. (The Central Singapore, SG), December 27, 2016.

118. C. Chua, *Standards, Quality Control, and Measurement Sciences in 3D Printing and Additive Manufacturing*, Academic Press, an imprint of Elsevier, London, UK, 2017.
119. ASTM International, Standard terminology for three-dimensional (3D) imaging systems, ASTM Standard ASTM E2544-11a, ASTM International, West Conshohocken, PA, 2011.
120. ASTM International, Standard practice for best practices for safe application of 3D imaging technology, ASTM Standard ASTM E2641-09, ASTM International, West Conshohocken, PA, 2017.
121. Technical Committee: ISO/IEC JTC 1/SC 29 , Printing material and 3D graphics coding for browsers conformance, ISO Standard ISO/IEC 14496-27, International Organization for Standardization, Geneva, Switzerland, 2009.
122. S. Singh and S. Ramakrishna, *Current Opinion in Biomedical Engineering*, Vol. 2, p. 105 , 2017.
123. J. Yuan, M. Zhu, B. Xu, and G. Chen, *Rapid Prototyping Journal*, Vol. 24, p. 409, 2018.
124. D. MacAdam, *Journal of the Optical Society of America*, Vol. 28, p. 466, 1938.
125. B.M. Rauzan, A.Z. Nelson, S.E. Lehman, R.H. Ewoldt, and R.G. Nuzzo, *Advanced Functional Materials*, Vol. 28, p. 1707032, 2018.
126. D. Safranski and J.C. Griffits, eds., *Shape-memory Polymer Device Design*, William Andrew, Oxford, 2017.
127. Q.P. Sun, R. Matsui, K. Tokida, and E.A. Pieczyska, eds., *Advances in Shape Memory Materials: In Commemoration of the Retirement of Professor Hisaaki Tobushi*, Springer, Cham, Switzerland, 2017.
128. Q. Zhao, H.J. Qi, and T. Xie, *Progress in Polymer Science*, Vol. 49-50, p. 79, 2015.
129. Z.X. Khoo, J.E.M. Teoh, Y. Liu, C.K. Chua, S. Yang, J. An, K.F. Leong, and W.Y. Yeong, *Virtual and Physical Prototyping*, Vol. 10, p. 103, 2015.
130. K. Yu, A. Ritchie, Y. Mao, M.L. Dunn, and H.J. Qi, *Procedia IUTAM*, Vol. 12, p. 193, 2015. IUTAM Symposium on Mechanics of Soft Active Materials.
131. Y.Y.C. Choong, S. Maleksaeedi, H. Eng, J. Wei, and P.-C. Su, *Materials & Design*, Vol. 126, p. 219, 2017.
132. M. Zarek, M. Layani, I. Cooperstein, E. Sachyani, D. Cohn, and S. Magdassi, *Advanced Materials*, Vol. 28, p. 4449, 2010.
133. C. Decker and A.D. Jenkins, *Macromolecules*, Vol. 18, p. 1241, 1985.
134. C.A.G. Rosales, M.F.G. Duarte, H. Kim, L. Chavez, D. Hodges, P. Mandal, Y. Lin, and T.-L. Tseng, *Materials Research Express*, Vol. 5, p. 065704, 2018.
135. D. Hua, X. Zhang, Z. Ji, C. Yan, B. Yu, Y. Li, X. Wang, and F. Zhou, *J. Mater. Chem. C*, Vol. 6, p. 2123, 2018.

136. R.U. Hassan, S. Jo, and J. Seok, *Journal of Applied Polymer Science*, Vol. 135, p. 45997, 2017.
137. Y. Mao, K. Yu, M.S. Isakov, J. Wu, M.L. Dunn, and H.J. Qi, *Scientific Reports*, Vol. 5, p. 13616, 2015.
138. Y. Mao, Z. Ding, C. Yuan, S. Ai, M. Isakov, J. Wu, T. Wang, M.L. Dunn, and H.J. Qi, *Scientific Reports*, Vol. 6, p. 24761, 2016.
139. C. Tingting, B. Hadi, L. Li, J. Jian, and A. Seema, *Advanced Functional Materials*, Vol. 28, p. 1870124, 2018.
140. C. Tingting, B. Hadi, L. Li, J. Jian, and A. Seema, *Advanced Functional Materials*, Vol. 28, p. 1800514, 2018.
141. J. Wu, C. Yuan, Z. Ding, M. Isakov, Y. Mao, T. Wang, M.L. Dunn, and H.J. Qi, *Scientific Reports*, Vol. 6, p. 24224, April 2016.
142. S. Shaffer, K. Yang, J. Vargas, M.A. Di Prima, and W. Voit, *Polymer*, Vol. 55, p. 5969, 2014.
143. N.A. Meisel, A.M. Elliott, and C.B. Williams, *Journal of Intelligent Material Systems and Structures*, Vol. 26, p. 1498, 2015.
144. R.C. Roberts, J. Wu, N.Y. Hau, Y.H. Chang, S.P. Feng, and D.C. Li, *Journal of Physics: Conference Series*, Vol. 557, p. 012006, 2014.
145. M. Zarek, M. Layani, S. Eliazar, N. Mansour, I. Cooperstein, E. Shukrun, A. Szlar, D. Cohn, and S. Magdassi, *Virtual and Physical Prototyping*, Vol. 11, p. 263, 2016.
146. S.K. Leist, D. Gao, R. Chiou, and J. Zhou, *Virtual and Physical Prototyping*, Vol. 12, p. 290, 2017.
147. H. Limei, J. Ruiqi, W. Jingjun, S. Jizhou, B. Hao, L. Bogeng, Z. Qian, and X. Tao, *Advanced Materials*, Vol. 29, p. 1605390, 2016.
148. Q. Ge, H.J. Qi, and M.L. Dunn, *Applied Physics Letters*, Vol. 103, p. 131901, 2013.
149. A. Sydney Gladman, E.A. Matsumoto, R.G. Nuzzo, L. Mahadevan, and J.A. Lewis, *Nature Materials*, Vol. 15, p. 413, January 2016.
150. S. Hirsch, A. Levy, and E. Napadensky, Water soluble ink-jet composition for 3D printing, US Patent 9 546 270, assigned to Stratasys Ltd. (Rehovot, IL), January 17, 2017.
151. K. Haubennestel, A. Bubat, D. Betcke, and J. Hartmann, Organosilane-modified polysiloxanes and their use for surface modification, US Patent 7 585 993, assigned to BYK-Chemie GmbH (DE), September 8, 2009.
152. H. Liu and C. He, Water-washable resin formulations for use with 3D printing systems and methods, US Patent 9 944 805, assigned to Full Spectrum Laser LLC, Las Vegas, NV, April 17, 2018.
153. I.E. Gunduz, M.S. McClain, P. Cattani, G.T.-C. Chiu, J.F. Rhoads, and S.F. Son, *Additive Manufacturing*, Vol. 22, p. 98, 2018.
154. E. Karjalainen, D.J. Wales, D.H.A.T. Gunasekera, J. Dupont, P. Licence, R.D. Wildman, and V. Sans, *ACS Sustainable Chemistry & Engineering*, Vol. 6, p. 3984, 2018.

155. J. Zhang and P. Xiao, *Polym. Chem.*, Vol. 9, p. 1530, 2018.
156. S.Y. Hong, Y.C. Kim, M. Wang, H.-I. Kim, D.-Y. Byun, J.-D. Nam, T.-W. Chou, P.M. Ajayan, L. Ci, and J. Suhr, *Polymer*, Vol. 145, p. 88, 2018.
157. A.T. DiBenedetto, *Journal of Polymer Science Part B: Polymer Physics*, Vol. 25, p. 1949, 1987.
158. J. Li, W.Y. Chan, K.K. Yee, and K.L.K. Cheuk, Photopolymer composition for 3D printing, US Patent 9 902 860, assigned to Nano and Advanced Materials Institute Limited (Hong Kong, CN), February 27, 2018.
159. H. Liu and C. He, Thermal and photo-initiation curing system of photopolymer resin for 3D printing, US Patent 9 777 097, assigned to Full Spectrum Laser LLC, October 3, 2017.
160. H. Liu and C. He, Photopolymer resin dual initiation wavelengths for 3D printing, US Patent 9 701 775, July 11, 2017.
161. S.K. Czaplewski, J. Kuczynski, J.T. Wertz, and J. Zhang, Dynamic polymer material for 3D printing, US Patent 9 944 826, assigned to International Business Machines Corporation (Armonk, NY), April 17, 2018.
162. G.G. Sacripante, K. Zhou, and T. Abukar, Copolymers for 3D printing, US Patent 9 738 752, assigned to Xerox Corporation (Norwalk, CT), August 22, 2017.
163. R.F. Zollo, *Cement and Concrete Composites*, Vol. 19, p. 107, 1997.
164. M. Hambach and D. Volkmer, *Cement and Concrete Composites*, Vol. 79, p. 62, 2017.
165. P. Colombo, G. Mera, R. Riedel, and G.D. Sorarù, *Journal of the American Ceramic Society*, Vol. 93, p. 1805.
166. Z.C. Eckel, C. Zhou, J.H. Martin, A.J. Jacobsen, W.B. Carter, and T.A. Schaedler, *Science*, Vol. 351, p. 58, 2016.
167. W. Easter and A. Hill, Additive manufacturing 3D printing of advanced ceramics, US Patent 9 944 021, assigned to Dynamic Material Systems LLC (Oviedo, FL), April 17, 2018.
168. S. Song, M. Park, J. Lee, and J. Yun, *Nanomaterials*, Vol. 8, p. 93, Feb 2018.
169. P. Colombo, J. Schmidt, G. Franchin, A. Zocca, and J. Günster, *American Ceramic Society Bulletin*, Vol. 96, p. 16, 2017.
170. A. Zocca, P. Colombo, C.M. Gomes, and J. Günster, *Journal of the American Ceramic Society*, Vol. 98, p. 1983, 2015.
171. G. Franchin, L. Wahl, and P. Colombo, *Journal of the American Ceramic Society*, Vol. 100, p. 4397, 2017.
172. Y. Fu, G. Xu, Z. Chen, C. Liu, D. Wang, and C. Lao, *Ceramics International*, Vol. 44, p. 11030 , 2018.
173. P. Jana, O. Santoliquido, A. Ortona, P. Colombo, and G.D. Sorarù, *Journal of the American Ceramic Society*, Vol. 101, p. 2732.

174. Z.C. Eckel, A.P. Nowak, A.M. Nelson, and A.R. Rodriguez, Monomer formulations and methods for 3D printing of preceramic polymers, US Patent Application 20 180 148 380, assigned to HRL Laboratories LLC, May 31, 2018.
175. J.W. Halloran, M. Griffith, and T.-M. Chu, Stereolithography resin for rapid prototyping of ceramics and metals, US Patent 6 117 612, assigned to Regents of the University of Michigan (Ann Arbor, MI), September 12, 2000.

3

Airplanes and Cars

An overview of 3D printing in manufacturing, aerospace, and the automotive industries has been given (1).

A critical analysis of additive manufacturing technologies for aerospace applications has been presented (2).

Emerging computational, experimental and processing innovations for advanced aircraft engines are expanding the scope for discovery and implementation of new metallic materials for future generations of advanced propulsion systems (3).

With additive manufacturing technologies it is possible to manufacture lightweight parts (4). In the automotive and aerospace industry the main goal is to make the lightest practical car or aircraft while securing safety.

Additive manufacturing technologies have enabled the manufacture of complex cross-sectional areas like the honeycomb cell (5) or every other material part that contains cavities and cutouts which reduce the weight-strength relation. It is possible to create lightweight structures; there are methods to get shapes that have a minimum weight, like the hanging method and the soap film method (6).

The hanging method and the soap film method produce a very difficult form of a structure, which has been used for civil construction, but with additive manufacturing it is possible to create structural parts for machines using the shape described by these methods and reducing the total weight. Selective laser sintering and electron beam melting are now used in the aircraft and aerospace industries. Engineers develop designs within the manufacturing constraints but this process expands the limits. With SLS and EBM, the limit will be the engineer's imagination. They open a whole new dimension of possible designs with almost any pre-alloyed metal powder (7).

In order to improve the hot-pressing process that is employed in laminated object manufacturing, an innovative heating-and-pressing separation system has been proposed, and heat transfer problems of this system were investigated. A thermal model was established. It is solved numerically by the finite element method, and could be verified by experiments. Based on the numerical solution under various operating conditions, it was found that the operating temperature of an adhesive can be reached quickly when the heater is maintained at a higher temperature, corresponding to a deeper heat-affected zone. This shortcoming can be effectively reduced if the speed of the heater is increased. Hence, a higher heater temperature together with a higher moving speed has been suggested to shorten processing time and promote manufacturing efficiency. Through analysis, the appropriate distance between the roller and the heater, so as to obtain finished parts of high quality, is determined (7).

With the traditional process these complex shape structures will be expensive to do if at all possible. With additive manufacturing printing technologies like selective laser sintering or electron beam melting, hollow structures, which are less expensive than solid ones, can be made since less material is used.

3.1 Airplanes

Aircraft applications are among the reviewed uses for additive manufacturing (8).

The results of a Navy workshop entitled "Direct Digital Manufacturing of Metallic Components: Affordable, Durable, and Structurally Efficient Aircraft"have been published (9). A vision of parts on demand when and where they are needed was articulated. Achieving the vision state would enhance operational readiness, reduce energy consumption, and reduce the total ownership cost of naval aircraft through the use of additive manufacturing. Specific technical challenges were identified to address the quantitative objectives in the areas of (8):

1. Innovative structural design,
2. Qualification and certification,
3. Maintenance and repair, and
4. Direct digital manufacturing science and technology.

The Atikins project concluded that an optimal design could show a weight and material saving of almost 40% (10). Their analysis showed that for long-range aircraft, reducing the weight of an aircraft by 100 *kg* results in both a 2.5 million dollar savings in fuel and a 1.3 *Mt* CO_2 savings over the lifetime of the aircraft.

3.1.1 *Material Testing Standards*

Material testing standards for additive manufacturing of polymer materials have been presented (11). These also include a building block approach for the support of composite structures in the 777 aircraft (12).

3.1.2 *Lightweight Aircraft Components*

The novel geometries enabled by additive manufacturing technologies can lead to performance and environmental benefits in a component's product application (13, 14). For example, the aircraft industry has adopted a number of different additive manufacturing components for reducing aircraft mass, including flight deck monitor arms, seat buckles, and various hinges and brackets, which can lead to greater aircraft fuel efficiency (15).

The life-cycle energy and greenhouse gas (GHG) emissions savings potential of additive manufacturing technologies for metallic aircraft components in the United States have been assessed (16).

3.1.3 *Aircraft Spare Parts*

Additive manufacturing is especially useful in the aircraft spare parts industry, where there is a need to maintain a high level of safety inventory for high-cost, long-lead time metallic parts (17).

Therefore, more and more companies in the aerospace industry are interested in using additive manufacturing technology. There are different approaches to configuring the aircraft spare parts supply chain using additive manufacturing technology.

This paper evaluates the impact of additive manufacturing in the aircraft spare parts supply chain based on the well-known supply chain operation reference model. Three supply chain scenarios are investigated; namely, conventional (as-is) supply chain, centralized additive manufacturing supply chain and distributed additive manufacturing supply chain. A case study is conducted based on data obtained in the literature. The result shows that the use of additive manufacturing will bring various opportunities for reducing the required safety inventory of aircraft spare parts in the supply chain. A sensitivity analysis is performed and some key factors affecting the choice of additive manufacturing scenarios are studied (17).

3.1.4 Polymer Laser Sintering

Laser sintering was introduced in the late 1990s as a manufacturing solution for end-use parts in the aerospace sector by Boeing and NAVAIR to supply low pressure ducting for the Boeing F/A-18 aircraft (18). However, the resulting supply chain specifications created a significant barrier to entry for the laser sintering process.

Commercial laser sintering machines have been used for years in a variety of rapid prototyping applications (18). Recently, the research in laser sintering has moved towards rapid manufacturing: The creation of engineered structural components.

However, the mechanical properties of laser sintering parts are often inconsistent compared to their molded counterparts. This occurs due to a variety of factors, including feedstock uniformity, microstructure evolution due to laser sintering processing, and the overall ability of commercial laser sintering machines to reliably form structural parts without thermal degradation of the feedstock powder.

The state-of-the-art of commercial laser sintering machines was reviewed. Also, the resulting implications for rapid manufacturing were discussed. Particular focus will be paid to the role of part bed temperature variations due to nonuniform heating, unsteady cooling due to natural convection currents in the part chamber, and how these and other phenomena may impact the design of laser control (18).

3.1.5 Composites Part Production

Four design principles have been shown that improve the production of composite parts during layup, handling, curing and post-processing in the layup process (19). Such design principles can be applied to a hat-stiffener, a highly integrated aircraft instrument panel and a novel insert eliminating drilling operations. The results indicate that additive manufacturing can reduce the part count, assembly steps and deformations during curing.

3.1.6 Deployable Wing Designs

The principles that are relevant to design for additive manufacturing have been presented (20). These include ideas about generating designs that cannot be fabricated using conventional methods to understanding the realities of existing machines and materials to microscale issues related to material microstructures and resulting process variations.

A topology optimization method has been developed for designing deployable structures with lattice-reinforced skins (21).

A deployable wing was designed for a miniature unmanned aerial vehicle (UAV). A physical prototype of the optimal configuration was fabricated with selective laser sintering and compared with the virtual prototype. The proposed methodology results in a 78% improvement in deviations from the intended surface profile of the deployed part (21).

UAVs have been developed to perform various military and civilian applications such as reconnaissance, attack missions, surveillance of pipelines, and interplanetary exploration (22).

There is a need to develop fast, adaptable UAV design technologies for agile, fuel-efficient, and flexible structures that are capable of adapting and operating in any environment (22).

An attempt has been made to develop adaptive design technologies by investigating current design methods and knowledge of deployable technologies in the area of engineering design and manufacturing. More specifically, this research seeks to identify one truss lattice with the optimal elastic performance for deployable UAV wing design according to the Hashin-Shtrikman theoretical bounds (23).

Three lattice designs have been proposed (22):

1. 3D Kagome structure,
2. 3D pyramidal structure, and
3. Hexagonal diamond structure.

The proposed lattice structure designs were fabricated using an Objet 350 3D printer while the material chosen was a poly(propylene)-like photopolymer called Objet DurusWhite RGD430 (24).

Based on compression testing, the proposed inflatable wing design can combine the advantages of compliant mechanisms and deployable structures to maximize flexibilities of movement in UAV design and development (22).

The selected photopolymer was a rubber-like material called Objet TangoGray FLX950 (22).

3.1.7 Additive Manufacturing for Aerospace

The philosophy of 3D printing in aerospace became: If you can design it, we can produce it (25). However, there is a laundry list of caveats with the current state of the technology, especially as it applies to aerospace applications and, in particular, launch vehicles.

The use of polymer-based materials has caught hold in production tooling, yet the use of polymers and metals for actual flight applications faces challenges until the technology catches up with the specific needs of the aerospace industry.

Perfection in quality, temperature extremes from cryogenic commodities to rocket engine exhaust, shock and vibration conditions, and the grand size need to be considered. However, given the variety of exquisite designs and low volume, aerospace applications are well suited for additive manufacturing. The challenges of today have already been solved and incorporated into this technology. It is simply a matter of time before additive manufacturing finds not only a foothold in aerospace but becomes the manufacturing tool of choice (25).

3.1.8 Fiber Reinforced Polymeric Components

Fiber reinforced polymeric components may be manufactured by coextruding a polymer filament and a fiber filament from a nozzle (26). The polymer filament and the fiber filament may be heated within the nozzle, which may couple the polymer filament to the fiber filament. The coextruded filament may result in long reinforcing fibers which provide strength to the components.

Many different types of components may be manufactured using these methods. For example, gas turbine engine aircraft components such as brackets, harness clips, inspection ports, hatch covers, etc., may be manufactured using these methods. The components may include an engine component, automotive component, sporting good component, consumer product component, tooling component, and assembly aid. The coextruded materials of components made by the methods described herein may allow for relatively lighter weight plastic components to replace components typically made of heavier materials such as steel or aluminum. The fiber filaments may provide added strength to the components even at relatively high temperatures, which can increase the strength of the polymer in order to be sufficient to replace heavier materials (26).

3.1.9 *Manufacturing of Aircraft Parts*

A method and an apparatus for manufacturing of aircraft parts has been presented (27).

Selective laser sintering may be used to build prototypes and production parts for use, such as in an aircraft. Selective laser sintering is capable of being used to produce parts with complex geometries within various dimensions.

Aircraft parts typically have stringent and/or extreme design requirements as compared to parts for other applications. These requirements may occur from operating environments that may have high loads and temperatures. Furthermore, these parts also may be required to be capable of withstanding impact loads from maintenance, handling, and/or other types of impact loads. For example, some parts may need to survive usage in some airframe locations that have in-service temperature ranges from around -54°C to around 225°C.

In particular, parts that exist near areas that are heated to or near engine or exhaust temperatures may need to be serviced and handled on the ground in severe winter conditions that may be present above 48 degrees north latitude or at altitude. These conditions require the material the parts are made of to have sufficient impact resistance at the low end of the temperature range. Simultaneously, sufficient stiffness and mechanical strength must be maintained at the high end of the temperature range to prevent failure in service.

A powder material is comprised of a polymer that is semicrystalline. The powder material has an overlap between a melting temperature range and a crystallization temperature range. The powder material also has a size particular distribution that is substantially a Gaussian distribution, a particle shape that is substantially spherical, and a desired melt flow rate that is less than a temperature at which the powder material begins to chemically break down (27).

3.1.10 *Multirotor Vehicles*

Flying drones, i.e., aircrafts that can fly autonomously, such as unmanned aerial vehicles (UAVs) or remote controlled piloted vehicles (RPVs), are now widely used and well established, although their use has so far been constrained by operational range due to their poor flight time (28).

Additive manufacturing has been selected to reduce the cost of the production of such prototypes to increase the manufacturing speed and to directly control the process parameters influencing the mechanical properties of the constructed parts. Moreover, the advantages related to the possibility of also building a small series or pre-series of those parts have been taken into account. The design process of such parts has been detailed (28).

3.1.11 *Flame Retardant Aircraft Carpet*

A lightweight carpet has been described for use in aircrafts that meets rigorous fire standards testing, is impervious to fluids, and is capable of being printed for decorative effect (29). The carpet may be composed of a layer of fire retardant treated poly(ethylene terephthalate) fibers adhered to a fire retardant treated poly(ethylene) (PE) film. The carpet is durable to normal foot traffic, resistant to most stains, non-fraying, and can be recycled.

The fibers of the face layer of the carpet may be pigmented to a mottled gray, for example, to provide a better background for subsequent printing, and/or the face layer fabric may be passed through a dye bath and dried to provide a gray background. The color may vary based on the subsequent design and colors in the printing process (29).

3.1.12 *Aircraft Cabins*

Poly(ether imide) (PEI) with the trade name Ultem 9085 is an amorphous and transparent polymer (30). This material is desirable due to its mechanical properties, relatively low density compared to traditional materials, and flame, smoke, and toxicity properties that allow its use in aircraft cabins. The PEI material is typically used on fused deposition modeling (FDM) machines for the manufacturing of end products (31).

3.1.13 *Additive Manufacturing of Solid Rocket Propellant Grains*

Solid rocket propellants are the workhorses in various strategic vehicles and space launch vehicles (32). Traditionally, composite propellant slurries are cast into cylindrical grains with one or more longitudinal ports to enable radial burning.

3.1.14 *High Temperature Heating System*

The amount of fiber reinforced composite material used in aircraft is continuously increasing. However, there is still limited application of composites for key load-carrying structural parts such as joints and lugs (33). The biggest challenge of fiber reinforced composites for key load-carrying parts is their geometric complexities, e.g., notches and holes, which undermine the structural benefit composites can bring about.

The goals of maintaining fiber continuity have been addressed so that load can be transferred along fibers and structural weight can be minimized. The application of 3D printing as a solution to manufacture complex structural parts has been identified as the means to achieve the proposed objective.

The design of a high temperature heating system has been described for a 3D printing application of continuous carbon fiber reinforced thermoplastic composites FDM process. The purpose of this system is to use poly(ether ether ketone) (PEEK) carbon fiber prepreg material as the printing material for continuous carbon fiber reinforced thermoplastic composites manufacturing.

The challenge for this project is to make a traditional FDM 3D printer capable of heating up the temperature to the PEEK melting point. The system includes a section for providing the heating, a section for temperature measurement and a closed-loop feedback proportional integral derivative control system (33).

As a semicrystalline thermoplastic, PEEK has excellent mechanical and chemical resistance properties at a much higher temperature than other thermoplastics. In the aerospace industry, PEEK material is considered one of the best materials for constructing airframe structures. The reasons for this include that it can provide significantly higher toughness and long-term resistance to fatigue and that it can be recycled more easily at the end of life, which makes the product life cycle more environmentally friendly.

Another significant benefit is that PEEK composites show particularly high resistance to fuel and hydraulic liquid corrosion, potentially reducing the maintenance requirements of the main rotor hub. For instance, the H160, a medium duty twin-engine civil helicopter made by Airbus, already has a rotor hub manufactured by PEEK resin matrix composites. This technology is designed by Porcher Industries, and its carbon fiber PEEK prepreg has already met the quality requirements of the safety critical application and received the green light for production by Airbus Helicopters (34).

3.1.15 Aerospace Propulsion Components

An overview of ongoing activities in additive manufacturing of aerospace propulsion components has been presented (35), which included rocket propulsion and gas turbine engines. Also, the opportunities concerning additive manufacturing of hybrid electric propulsion components were discussed.

In particular, the 3D printing of a small gas turbine engine and 3D printing of cooled turbine engine blades have been shown (35). The fabrication time was reduced from 1 *y* to 4 months with 70% less cost. Also, additive manufacturing of a 3D printed rocket engine turbo pump has been demonstrated.

Nonmetallic gas turbine engine components are shown in Table 3.1.

Table 3.1 Nonmetallic gas turbine engine components (35).

Component	Material
Fan duct	Polymer
Exhaust Components	Ceramic matrix composite
Turbine shroud & nozzles	Ceramic matrix composite
Fan bypass Stator	Polymer matrix composite
Compressor vanes	Polymer matrix composite
Engine access panel and acoustic liner	Polymer

The fabrication of high temperature polymer matrix composite was enabled by: Chopped fiber reinforcement, moisture reduction in the FDM filament, and a versatile printing pattern design.

Also, fused deposition modeling simplifies the fabrication of an acoustic liner.

An integral facesheet/honeycomb structure can be fabricated in one step using fused deposition modeling standard liner configuration complex geometries. The device is fabricated from a monolithic Ultem 9085 thermoplastic with a glass transition temperature of 186°C (367°F).

3.1.16 Antenna RF Boxes

The Boeing Company has used additive manufacturing to manufacture low volume parts for F18 fighter jets and 787 jet airliners, as well as having exploited the design opportunities for part consolidation, reducing the number of pieces in one design from 15 to 1 (36). The most prevalent application of additive manufacturing has been in the hearing aid industry.

Saab Avitronics has used additive manufacturing to produce antenna RF boxes for unmanned aircraft that are more compact, have a 45% mass reduction and contain complex features that have traditionally been very difficult to manufacture using conventional techniques (37)

There are also other known examples in a variety of sectors (38).

3.1.17 *Cyanate Ester Clay Nanocomposites*

An improvement in aircraft cabin material fire safety has been a much desired objective (39). In the pursuit of improved approaches to the design of ultra-fire-resistant aircraft interior materials, a wide variety of concerns must be addressed in addition to flammability.

For many polymers used in the interior of aircraft, cost is a major concern for aircraft manufacturers. Therefore, an additive approach to improving the fire safety of polymers is attractive. The additives must be inexpensive and easily processed with the polymer. In addition, the additive must not excessively degrade the other performance properties of the polymer, and it must not create environmental problems when recycling or at the time of its final disposal.

It has been found that polymer layered-silicate (clay) nanocomposites have the unique combination of improved flammability properties and improved physical properties as well as recyclability.

The use of a layered silicate, montmorillonite clay has been reported, dispersed at the nanometer level, in cyanate ester resins for improving flammability. It was shown that the use of melamine-treated montmorillonite in these resins yields exfoliated montmorillonite in the cured cyanate ester nanocomposites. This reduces the peak heat release rate by more than 50%.

This nanocomposite approach would be especially useful in improving toughened cyanate ester resins since the typical toughening agents used often increase the flammability and lower the modulus (39).

3.1.18 *Bionic Lightweight Design*

A bionic lightweight design for the aircraft industry has been reported (40, 41). The fabrication has been done by laser additive manufacturing.

The individual production steps of the laser additive manufacturing process are as follows: In the preprocessing step, a 3D-CAD model is divided into horizontal slices with thickness corresponding to the layers in the production process. Typical layer thicknesses for $TiAl_6V_4$ are in the range of 30 μm to 50 μm. Subsequent to the slicing, the prepared data are transmitted to the selective laser melting machine, where the actual manufacturing process occurs in three repeating steps. First a powder layer is applied to the base plate and in the second step it is exposed to the laser beam. Due to the energy input of the focused laser beam, the powder starts melting and solidifies into the welding beads after exposure. In every single layer the beam melts the surface area corresponding to the CAD model slice. With each new layer, the base plate is lowered and powder is applied. After completion of all the layers the part can be taken from the powder bed and unmelted powder can be recycled for further production (40).

The above described stepwise production causes the reduction of complex three-dimensional geometries into simple two-dimensional manufacturing steps and enables the production of highly complex parts. Thus, nearly any structure can be manufactured, which facilitates the realization of designs in accordance with the parts' stress distributions under load. Therefore, this method allows new design approaches for lightweight structures that were not possible up to now, due to the restrictions of conventional manufacturing processes.

The development of the material used therein is based on an analytical calculation of temperature distribution versus effective process factors in order to identify acceptable operating conditions for the laser additive manufacturing process.

An approach for an extreme lightweight design was realized by incorporating structural optimization tools and bionic structures into the process. As a consequence of following these design principles, lightweight savings can be achieved for designing new aircraft structures (40).

Also, design guidelines for laser additive manufacturing of lightweight structures in $TiAl_6V_4$ have been reported (42). Thin walls, bars, and bore holes with varying diameters were built in different orientations to determine the process limits. From the results of the experiments, comprehensive design guidelines for lightweight structures were derived in a catalog according to DIN 2222 (43) and were presented in detail. For each structure, a favorable and unfavorable example was shown, and the underlying process restrictions were mentioned.

3.2 Cars

The reasons as to why 3D printing has had such a large impact on the automotive industry and how it can be implemented to improve performance while lowering lead times and cost have been discussed (44).

For the automotive industry, recent advances in additive manufacturing have opened doors for newer, more robust designs, resulting in (45):

1. Lighter, stronger, and safer products,
2. Reduced lead times, and
3. Reduced costs.

In 2015, the annual Wohlers report (46, 47) stated that the automotive industry accounted for 16.1% of all additive manufacturing expenditures (44).

While automotive original equipment manufacturers and suppliers primarily use additive manufacturing for rapid prototyping, the technical trajectory of additive manufacturing makes a strong case for its use in product innovation and direct manufacturing in the future (44).

In this article, the design requirements have been discussed for parts used in vehicles as well as design recommendations for common automotive applications. A range of popular additive manufacturing materials suitable for the automotive industry are presented along with several case studies where additive manufacturing has successfully been implemented.

Since the production volume in the automotive industry is generally very high (greater than 100,000 parts per year), additive manufacturing has predominantly been used as a prototyping solution rather than for end part manufacturing. Improvements in the size of industrial printers, the speed they are able to print at and the materials that are available mean that additive manufacturing is now a viable option for many medium-sized production runs, particularly for higher-end automobile manufacturers that restrict production numbers to far fewer than the average.

One of the most critical aspects relating to the automotive industry is the weight reduction of components. Automotive applications make use of advanced engineering materials and complex geometries in an attempt to reduce weight and improve performance. Additive manufacturing is capable of producing parts from many of the lightweight polymers and metals that are common in the automotive industry. Certain uses are shown in Table 3.2.

Table 3.2 Applications in the automotive industry (44).

Application	Process	Material	Features and Examples
Under the hood	SLS	Nylon	Heat resistant functional parts, Battery cover
Interior accessories	SLA	Resin	Customized cosmetic components, Console prototype
Air ducts	SLS	Nylon	Flexible ducting and bellows, Air conditioning ducting
Full-scale panels	SLA	Resin	Large parts for sanding and painting, front bumper
Cast metal brackets	SLA	Wax	Metal parts, Alternator mounting bracket
Complex metal components	DMLS	Metal	Consolidated, lightweight, functional metal parts, Suspension wishbone
Bezels	Jetting	Photopolymer	End use custom screen bezels, Dashboard interface
Lights	SLA	Resin	Fully transparent, high detail models, Headlight prototypes

3.2.1 *Laser Sintering*

In recent years, PE, poly(urethane) and other thermoplastic elastomeric materials have been introduced for laser sintering to fabricate flexible and smart actuators or dampers for wearable devices and automotive applications (48–51).

The engineered parts for automotive application are an intake manifold and inlet manifold (52)

3.2.2 *Automotive Repair Systems*

In many cases, if only a portion of an automotive part is damaged, the repair of that portion requires a replacement of the entire part (53, 54). As a result, automotive repair processes with respect to the automotive part are not flexible, requiring the same replacement procedure to be followed regardless of the degree to which the part is damaged.

Moreover, such a replacement procedure is expensive because the entire automotive part must be purchased. The replacement procedure may also be very time-consuming, especially when the entire automotive part must be ordered or otherwise obtained from an external source.

An automotive repair system has been described (54) that may be used for repairing automotive parts, including methods or systems that employ 3D printing or three-dimensionally printed parts such as, for example, 3D printed portions of an automotive part, connector tabs, attachment parts, etc. The automotive repair system includes a hand-held device and a controller.

3.2.3 *Improving Aerodynamic Shapes*

Aerodynamics is a division of fluid dynamics concerned with the flow of air (55). Understanding the flow of air around a body allows the calculation of forces and moments acting on the object.

The computer-aided design models of sedan bodies were simulated and analyzed with regard to their aerodynamics, especially on the drag and lift estimation. In addition, computational fluid dynamics (CFD) analysis was developed as a possible procedure for drag estimation and aerodynamics studies on the body. The printed models were examined in a wind tunnel test (55).

3.2.4 *Common Automotive Applications*

Selective laser sintering (SLS) can be used to make semi-functional bellow pieces, where some flexibility is required in assembly or mating. In general, this material/process is considered to be best for applications where the part will be exposed to very few repetitive flexing motions. For projects that require significant flexing, other poly(ethylene)-based SLS materials, such as DuraForm Flex, are better suited (44).

3.2.4.1 *Functional Mounting Brackets*

Being able to rapidly manufacture a complex, lightweight bracket overnight is a trademark of the additive manufacturing industry. Not only does additive manufacturing allow for organic shapes and designs to be manufactured but it also requires very little input from an operator, meaning that engineers are able to quickly take a design from computer to assembly in a very short amount of time. This is not possible with traditional manufacturing techniques, like computerized numerical control machining, where a highly skilled machine operator is needed to produce parts. Powder bed fusion technologies, like SLS nylon and metal printing, are best suited for functional parts and offer a range of materials (from PA12 nylon to titanium) (44).

3.2.4.2 *Deformable Armrest*

An armrest is a feature routinely added to modern vehicles to provide comfort to the driver and passengers who may wish to rest their arms while sitting in the vehicle (56). Several different types of armrests have been developed, including an armrest built into a vehicle door or back panel.

An automotive armrest has been described (56). The armrest includes an elastic core defining a patterned array of channels having the same orientation extending throughout. The elastic core exhibits a predetermined target vertical stiffness and a predetermined target lateral stiffness different than the predetermined target vertical stiffness. The armrest further includes a skin arranged with the elastic core to form the automotive armrest. Each of the channels has a cross section defined by a lateral dimension and a vertical dimension. The lateral dimension is less than the vertical dimension. The cross sections may be ellipses or rhombuses. The channels may be arranged in a regular pattern. The widths, lengths, or both of some of the channels may be different. The channels may extend along a vertical axis of the core. The patterned array may include a same number of the channels in each row, column, or both. The elastic core and skin may be 3D printed.

The type of the material used for 3D printing determines the type of the 3D printer to be utilized. For example, if a resin is used, stereolithography, digital light processing, or a multijet 3D printer system may be implemented. Alternatively, if polymeric materials are used, fused deposition modeling or selective laser sintering may be viable methods of producing the armrest.

Exemplary materials for the core or the skin may be 3D printed from acrylonitrile-butadiene-styrene (ABS), and ABS-based thermoplastics poly(carbonate), a PEI resin, poly(phenylsulfone), nylon, poly(amide), poly(styrene), poly(lactic acid), and others. Other exemplary materials may include liquid UV curable photopolymer resins such as epoxy-based or acrylate-based resins (56).

3.2.4.3 *Vehicular Brake Components*

A method for making a vehicular brake component has been described. The method consists of (57):

1. Providing a three-dimensional printer,
2. Providing the printer with a schematic for making a preform brake rotor or hub,
3. Supplying a metal powder to the printer for making the preform brake rotor or hub,
4. Forming the preform brake rotor or hub per the schematic provided and the metal powder supplied to the printer,
5. Sintering the preform brake rotor or hub, and
6. Applying a wear coating to the sintered preform brake rotor or hub to make the brake component therefrom.

Preferably, such brake components for automotive racing parts are made from titanium alloy powders (57).

3.2.4.4 *Drives*

In July 2016, a Daimler press release described the use of 3D printing by Mercedes-Benz Trucks, part of the Daimler Trucks division, to generate spare parts (58).

Mercedes-Benz Trucks was using the latest 3D printing processes for plastic spare parts as the standard production method in the customer services and parts sector. At Daimler, more than 100,000 printed prototype parts were manufactured for the individual company divisions every year.

A thumb cast for assembly line workers was developed by BMW (59). The workers had to push a huge number of rubber plugs into holes on the assembly line, which was giving too many of them a repetitive strain type of injury. A cast of the thumb and hand took all the strain out of the process.

Even more revolutionary proposals have been pursued. BMW and Erik Melldahl, a Swedish designer, have proposed a 3D printed concept car with a completely degradable main body made from a mixture of mycelium mushrooms and grass (60). The goal of the project was to explore some ways to produce a sustainable, locally produced car, even in areas such as Africa's Serengeti (58).

3.2.4.5 3D Printed Vehicle

The world's first 3D printed vehicle emerged during the 2014 International Manufacturing Technology Show (61). In a matter of two days, history was made at Chicago's McCormick Place, when the world's first 3D printed electric car, named Strati, the Italian world for *layers*, was presented.

The numerical simulation of big area 3D printing of a full-size car has been described (62). This is based on the finite element analysis method. The finite element analysis-based solution is specifically implemented on the big area additive manufacturing process of an entire car.

In cooperation with Cincinatti Inc. and Local Motors, ORNL has implemented the big area additive manufacturing process to produce the Strati car. The big area additive manufacturing equipment consists of a pellet dryer, an air-driven pellet conveying system that supplies dry pellets into a hopper connected to a single screw extruder, which is mounted vertically on a precision XYZ gantry. The gantry and extruder is digitally controlled. The part is deposited onto a heated bed, whose vertical motion is also synchronized with the gantry and the extruder (62).

3.2.5 *Thermomechanical Pulp Fibers*

The use of natural fiber-based biocomposites presents an opportunity for applications in the automotive industry, where lightweight construction is an important factor. At the present, automotive companies are attempting to reduce the use of man-made fibers by the addition of natural fibers in the non-structural plastic parts of vehicles. Biocomposites from biobased matrices like natural fibers are emerging as new materials (63).

However, natural fibers have some limitations, e.g., processing temperatures below 220°C and a possible incompatibility between hydrophilic fibers and commonly hydrophobic polymer matrices (64).

The use of maleated poly(ethylene) as coupling agent is an attractive method because it avoids the use of expensive and toxic reagents (65). The improvement of the interfacial adhesion between fibers and matrix may lead to improved physical and mechanical properties of the biocomposites.

Biobased PE and thermomechanical pulp (TMP) fibers were used to produce biocomposites (66).

Biobased PE is a chemically identical alternative to PE from a petrochemical feedstock.

Since PE is a semicrystalline polymer, it shrinks and warps during solidification. Such a characteristic may negatively affect the 3D performance of PE.

However, a thermomechanical pulp reinforced biobased PE can be manufactured with an adequate mechanical performance for injection molding. Additionally, it was demonstrated that such biocomposites are suitable for manufacturing printable filaments for fused deposition modeling (66).

3.2.6 *Polyamic Acid Salts*

Polyamic acid salts are suitable for use in the photocuring additive manufacturing processes of all-aromatic poly(imide)s (PIs) (67). Due to the all-aromatic structure, these high-performance polymers are exceptionally chemically and thermally stable but are not conventionally processable in their imidized form.

The facile addition of 2-(dimethylamino)ethyl methacrylate to commercially available poly(4,4′-oxydiphenylene pyromellitamic acid) afforded ultraviolet curable polyamic acid salt solutions. These readily prepared solutions do not require multistep synthesis, exhibited fast gel times of less than 5 *s*, and rendered high G′ gel-state moduli. Vat photopolymerization 3D printing afforded self-supporting organogels.

A subsequent thermal treatment rendered the crosslinked polyamic acid precursor to an all-aromatic PI. This fast and facile strategy makes such composites accessible in three dimensions, offering a way to impact aerospace and automotive technologies (67).

3.2.7 *Recycled Tempered Glass from the Automotive Industry*

Additive manufacturing has been applied in many industries to produce automotive, aerospace, medical and other commercial products. The technologies are supported by materials for the manufacturing process to produce high quality products. Nowadays, the high technology of additive manufacturing really needs high investment to carry out the process for fine products. Polymer, metal and ceramic are the three foremost types of materials used for additive manufacturing application, which were mostly in the form of wire feedstock or powder. In this circumstance, it is crucial to recognize the characteristics of each type of material used in order to understand the behavior of the materials in high temperature application via additive manufacturing. A review has been presented that aims to provide excessive investigation and gather the necessary information for further research on additive materials for high temperature application (68).

Also, a new material has been proposed that comes from recycled tempered glass from the automotive industry, which has huge potential for high temperature application. The technique proposed for additive manufacturing will minimize some cost of modeling with the same quality of products compared to other advanced technologies used for high temperature application (68).

References

1. C.W.J. Lim, K.Q. Le, Q. Lu, and C.H. Wong, *IEEE Potentials*, Vol. 35, p. 18, July 2016.
2. A. Angrish, A critical analysis of additive manufacturing technologies for aerospace applications, in *2014 IEEE Aerospace Conference*, pp. 1–6, March 2014.
3. T.M. Pollock, *Nature Materials*, Vol. 15, p. 809, July 2016.
4. K.V. Wong and A. Hernandez, *ISRN Mechanical Engineering*, Vol. 2012, p. 1, 2012.
5. K.-U. Bletzinger and E. Ramm, *Computers & Structures*, Vol. 79, p. 2053, 2001.
6. A. Williams, *Architectural Research Quarterly*, Vol. 6, p. 337, 2002.
7. Y. Liao, H. Li, and Y. Chiu, *The International Journal of Advanced Manufacturing Technology*, Vol. 27, p. 703, Jan 2006.
8. W.E. Frazier, *Journal of Materials Engineering and Performance*, Vol. 23, p. 1917, Jun 2014.
9. W.E. Frazier, Digital manufacturing of metallic components: Vision and roadmap, in *Solid Free Form Fabrication Proceedings*, pp. 717–732, Austin, TX, 2010. University of Texas at Austin.
10. G. Hague, *Southborough University, UK*, 2010.
11. A.M. Forster, Materials testing standards for additive manufacturing of polymer materials, electronic: https://nvlpubs.nist.gov/nistpubs/ir/2015/NIST.IR.8059.pdf, 2015.
12. Building block approach for composite structures in S.T. Peters, ed., *Composite Materials Handbook-MIL 17: Materials Handbook: Polymer matrix composites materials usage, design, and analysis*, Vol. 1, chapter 4, pp. 24–29. American Society for Testing and Materials, 3rd edition, 2002.
13. T.J. Horn and O.L.A. Harrysson, *Science Progress*, Vol. 95, p. 255, 2012.
14. S.H. Huang, P. Liu, A. Mokasdar, and L. Hou, *The International Journal of Advanced Manufacturing Technology*, Vol. 67, p. 1191, 2013.
15. J.-P. Immarigeon, R.T. Holt, A.K. Koul, L. Zhao, W. Wallace, and J.C. Beddoes, *Materials Characterization*, Vol. 35, p. 41, 1995.
16. R. Huang, M. Riddle, D. Graziano, J. Warren, S. Das, S. Nimbalkar, J. Cresko, and E. Masanet, *Journal of Cleaner Production*, Vol. 135, p. 1559, 2016.
17. P. Liu, S.H. Huang, A. Mokasdar, H. Zhou, and L. Hou, *Production Planning & Control*, Vol. 25, p. 1169, 2014.
18. D.L. Bourell, T.J. Watt, D.K. Leigh, and B. Fulcher, *Physics Procedia*, Vol. 56, p. 147, 2014. 8th International Conference on Laser Assisted Net Shape Engineering LANE 2014.

19. D.-A. Türk, R. Kussmaul, M. Zogg, C. Klahn, B. Leutenecker-Twelsiek, and M. Meboldt, *Procedia CIRP*, Vol. 66, p. 306 , 2017. 1st CIRP Conference on Composite Materials Parts Manufacturing (CIRP CCMPM 2017).
20. D.W. Rosen, *Virtual and Physical Prototyping*, Vol. 9, p. 225, 2014.
21. U.M. Namasivayam and C.C. Seepersad, *Rapid Prototyping Journal*, Vol. 17, p. 5, 2011.
22. S.K. Moon, Y.E. Tan, J. Hwang, and Y.-J. Yoon, *International Journal of Precision Engineering and Manufacturing-Green Technology*, Vol. 1, p. 223, July 2014.
23. Z. Hashin and S. Shtrikman, *Journal of the Mechanics and Physics of Solids*, Vol. 11, p. 127, 1963.
24. N. N., Durus white (rgd430), electronic: https://www.3dhubs.com/material/stratasys-durus-white-rgd-430, 2018.
25. G.J. Schiller, Additive manufacturing for aerospace, in *2015 IEEE Aerospace Conference*, pp. 1–8, March 2015.
26. E. Butcher, W. Twelves, G. Schirtzinger, J. Ott, and L. Dautova, Methods for manufacturing fiber-reinforced polymeric components, US Patent 9 931 776, assigned to United Technologies Corporation (Farmington, CT), April 3, 2018.
27. C.S. Huskamp and B.I. Lyons, Manufacturing aircraft parts, US Patent 8 709 330, assigned to The Boeing Company (Chicago, IL), April 29, 2014.
28. C. Ferro, R. Grassi, C. Seclì, and P. Maggiore, *Procedia CIRP*, Vol. 41, p. 1004, 2016.
29. S.W. Foss and C. Deerney, Fire-retardant, lightweight aircraft carpet, US Patent Application 20 060 240 217, assigned to Foss Manufacturing Co., Inc., Hampton (NH), October 26, 2006.
30. W. Hellerich, G. Harsch, and S. Haenle, *Werkstoff-Führer Kunststoffe: Eigenschaften, Prüfungen, Kennwerte*, Carl Hanser Verlag, München, 8th edition, 2004.
31. A. Bagsik, V. Schöppner, and E. Klemp, FDM part quality manufactured with Ultem 9085, in *14th International Scientific Conference on Polymeric Materials*, Vol. 15, pp. 307–315, 2010.
32. R.A. Chandru, N. Balasubramanian, C. Oommen, and B.N. Raghunandan, *Journal of Propulsion and Power*, pp. 1–4, 2018.
33. J. Pu, P. Zhuo, A. Jones, S. Li, I. Ashcroft, T. Yang, and K. Ponggorn, Temperature control of continuous carbon fibre reinforced thermoplastic composites by 3D printing, in *21st International Conference on Composite Materials*, number 436, pp. 1–7, 2017.
34. N. N., Airbus helicopters h160 – a new chapter in helicopter design, electronic: http://pipreg.com/medias/dossier-presse/pdf/porcher-dossier-presse_11.pdf, 2017.

35. A.K. Misra, J.E. Grady, and R. Carter, Additive manufacturing of aerospace propulsion components, electronic: https://ntrs.nasa.gov/search.jsp?R=20150023067, 2015.
36. G.N. Levy, R. Schindel, and J. Kruth, *CIRP Annals*, Vol. 52, p. 589, 2003.
37. R.D. Goodridge, C.J. Tuck, and R.J.M. Hague, *Progress in Materials Science*, Vol. 57, p. 229, 2012.
38. P. du Plessis, S. Avitronics, and R. Arena, A functional application of RM in a military environment, in *TCT Conference*, pp. 1–13. Ricoh Arena Coventry, 2008.
39. J.W. Gilman, J. Harris, Richard H., and D. Hunter, Cyanate ester clay nanocomposites: synthesis and flammability studies, in *Evolving and Revolutionary Technologies for the New Millennium*, Long Beach, CA, 1999. NIST.
40. C. Emmelmann, M. Petersen, J. Kranz, and E. Wycisk, Bionic lightweight design by laser additive manufacturing (LAM) for aircraft industry, in *Sustainable Design, Manufacturing, and Engineering Workforce Education for a Green Future*, Vol. 8065, pp. 1–12, Strasbourg, France, 2011. SPIE Eco-Photonics 2011.
41. C. Emmelmann, P. Sander, J. Kranz, and E. Wycisk, *Physics Procedia*, Vol. 12, p. 364, 2011. Lasers in Manufacturing 2011 - Proceedings of the Sixth International WLT Conference on Lasers in Manufacturing.
42. J. Kranz, D. Herzog, and C. Emmelmann, *Journal of Laser Applications*, Vol. 27, p. S14001, 2015.
43. DIN Standard, Taper ring gauges for taper shafts for drill chucks, DIN Standard 222, Deutsches Institut für Normung, Berlin, 1982.
44. B. Artley, Automotive 3D printing applications, electronic: https://www.3dhubs.com/knowledge-base/automotive-3d-printing-applications, 2018.
45. C.A. Giffi, B. Gangula, and P. Illinda, 3D opportunity in the automotive industry. additive manufacturing hits the road, electronic: https://www2.deloitte.com/content/dam/insights/us/articles/additive-manufacturing-3d-opportunity-in-automotive/DUP_707-3D-Opportunity-Auto-Industry_MASTER.pdf, 2014.
46. T. Wohlers, *Wohlers Report 2015: 3D Printing and Additive Manufacturing State of the Industry, Annual Worldwide Progress Report*, Wohlers Associates, Fort Collins, Colorado, 2015.
47. T. Wohlers, *Wohlers Report 2017: 3D Printing and Additive Manufacturing State of the Industry: Annual Worldwide Progress Report*, Wohlers Associates, Fort Collins, Colorado, 2017.
48. F. Shen, S. Yuan, Y. Guo, B. Zhao, J. Bai, M. Qwamizadeh, C.K. Chua, J. Wei, and K. Zhou, *International Journal of Applied Mechanics*, Vol. 8, p. 1640006, 2016.

49. S. Yuan, J. Bai, C. Kai Chua, K. Zhou, and J. Wei, *Journal of Computing and Information Science in Engineering*, Vol. 16, p. 041007, November 2016.
50. S. Yuan, C.K. Chua, and K. Zhou, Dynamic mechanical behaviors of laser sintered polyurethane incorporated with MWCNTs, in *Proceedings of the 2nd International Conference on Progress in Additive Manufacturing (Pro-AM 2016)*, pp. 361–366, Singapore, 2016. Research Publishing, Singapore.
51. S. Yuan, F. Shen, J. Bai, C.K. Chua, J. Wei, and K. Zhou, *Materials & Design*, Vol. 120, p. 317, 2017.
52. S. Yuan, Y. Zheng, C.K. Chua, Q. Yan, and K. Zhou, *Composites Part A: Applied Science and Manufacturing*, Vol. 105, p. 203, 2018.
53. D. Harrier and J. Milazzo, Automotive repair systems including three-dimensional (3D) printed attachment parts and methods of use, US Patent Application 20 170 368 770, assigned to Service King Paint & Body, LLC, December 28, 2017.
54. D. Harrier, Automotive repair systems including three-dimensional (3D) printed attachment parts and methods of use, WO Patent 2 017 223 375, assigned to Service King Paint & Body, LLC, December 28, 2017.
55. P. Shakti, P.N. Bhatt, R.G. Gill, A.B. Jadeja, and D.J. Parekh, *IJNRD*, Vol. 3, p. 82, 2018.
56. M.O. Faruque, J.C. Cheng, and Y. Chen, Deformable armrest having a patterned array of channels, US Patent 9 994 137, assigned to Ford Global Technologies, LLC (Dearborn, MI), June 12, 2018.
57. G. Martino, Method for making vehicular brake components by 3D printing, US Patent Application 20 180 093 414, April 5, 2018.
58. T. Lecklider, *Evaluation Engineering*, Vol. 21, 2016.
59. N. Hall, *3D Printing Industry*, Vol. 20, 2016.
60. H. Milkert, BMW introduces future 3D printable concept vehicle specifically designed for the Serengiti, electronic: https://3dprint.com/9174/bmw-3d-printed-vehicle/, 2014.
61. M. Chinthavali, 3D printing technology for automotive applications, in *2016 International Symposium on 3D Power Electronics Integration and Manufacturing (3D-PEIM)*, pp. 1–13, June 2016.
62. M.R. Talagani, S. DorMohammadi, R. Dutton, C. Godines, H. Baid, F. Abdi, V. Kunc, B. Compton, S. Simunovic, and C. Duty, *SAMPE Journal*, Vol. 51, p. 27, 2015.
63. S. Siengchin, *Express Polymer Letters*, Vol. 11, p. 600, 2017.
64. K.L. Pickering, M.G.A. Efendy, and T.M. Le, *Composites Part A: Applied Science and Manufacturing*, Vol. 83, p. 98, 2016. Special Issue on Biocomposites.
65. N.C. Liu and W.E. Baker, *Polymeric Materials Science and Engineering*, Vol. 11, p. 249, 1992.

66. Q. Tarrés, J. Melbø, M. Delgado-Aguilar, F. Espinach, P. Mutjé, and G. Chinga-Carrasco, *Composites Part B: Engineering*, Vol. 153, p. 70, 2018.
67. J. Herzberger, V. Meenakshisundaram, C.B. Williams, and T.E. Long, *ACS Macro Letters*, Vol. 7, p. 493, 2018.
68. N.A.B. Nordin, M.A. Bin Johar, M.H.I. Bin Ibrahim, and O.M.F. Bin Marwah, *IOP Conference Series: Materials Science and Engineering*, Vol. 226, p. 012176, 2017.

4

Electric and Magnetic Uses

4.1 Electric Uses

4.1.1 Conductive Microstructures

A 3D printing system based on liquid deposition modeling has been developed for the fabrication of conductive 3D nanocomposite-based microstructures with arbitrary shapes.

The technology consists of the additive multilayer deposition of polymeric nanocomposite liquid dispersions based on poly(lactic acid) (PLA) and multiwalled carbon nanotubes (CNTs) with a home-modified, low-cost commercial benchtop 3D printer.

Measurements of the electrical and rheological properties at varying concentrations of CNT and PLA were used to find the optimal processing conditions and the printability windows for these systems.

The electrical conductivity as a function of the CNT concentration in PLA is shown in Figure 4.1.

As shown in Figure 4.1, the conductivity is found to already increase substantially upon the addition of 0.5% CNTs. In addition, a progressive increase in the electrical conductivity is observed for increasing CNT concentrations following a typical percolation behavior, until values in the range of 10–100 $S\,m^{-1}$ are reached for highly concentrated CNT/PLA nanocomposites in the range of 5% to 10%.

A uniform homogeneous dispersion of the CNTs within a polymeric matrix is a key factor for the development of printable CNT-based nanocomposites, because of the high tendency of the

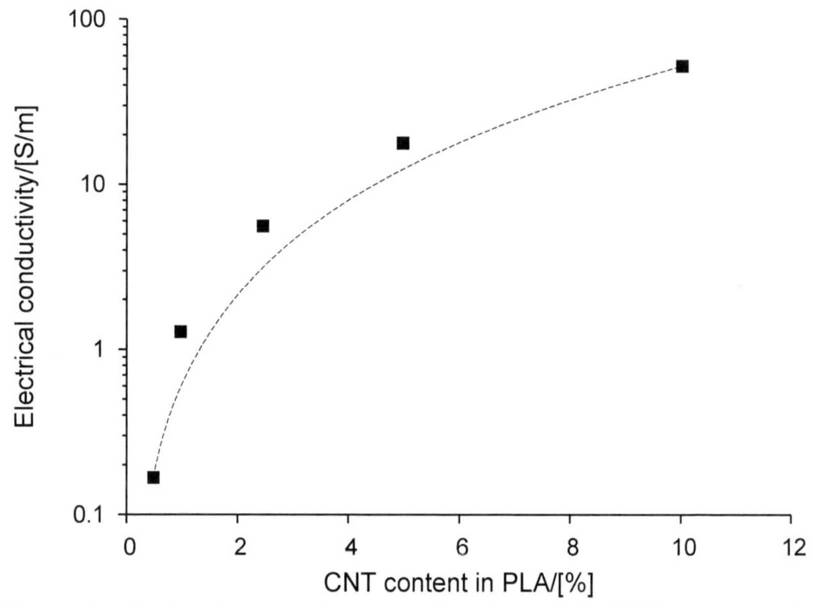

Figure 4.1 Electrical conductivity as a function of the CNT concentration (1).

CNTs to form bundles and aggregates that may cause clogging of the printing nozzle and flux instability during the printing process (2). Therefore, an appropriate CNT concentration is required to obtain a nanocomposite material that is simultaneously conductive and 3D printable (1).

The results of this study open the way to the direct deposition of intrinsically conductive polymer-based 3D microstructures by means of a low-cost liquid deposition modeling 3D printing technique (1).

4.1.1.1 Core-Shell Nanowires

A synthesis method has been reported that yields 4.4 *g* of Cu nanowires in 1 *h*, and a method to coat 22 *g* of Cu nanowires with Ag within 1 *h* (3).

Due to the large diameters of Cu nanowires (around 240 *nm*) produced by this synthesis, a Ag:Cu mol ratio of 0.04 is sufficient to coat the nanowires with ca. 3 *nm* of Ag, and thereby protect them from oxidation. This multigram Cu-Ag core-shell nanowire production process enabled the production of the first nanowire based conductive polymer composite filament for 3D printing. The 3D printing filament has a resistivity of 0.002 $\Omega\, cm$, so is greater than 100 times more conductive than commercially available graphene-based 3D printing filaments.

The conductivity of composites containing 5 vol% of 50 μm long Cu-Ag nanowires is greater than composites containing 22 vol% of 20 μm long Ag nanowires or 10 μm long flakes, indicating that high-aspect-ratio Cu-Ag nanowires enable the production of highly conductive composites at relatively low volume fractions. The highly conductive filament can support current densities between 2.5 and 4.5 $10^{-5}\ Am^{-2}$ depending on the surface-to-volume ratio of the printed trace, and was used to 3D print a conductive coil for wireless power transfer (3).

4.1.1.2 *Electrically Conductive Polymer Nanocomposites*

Acrylonitrile-butadiene-styrene (ABS) is chosen for the compositions as the matrix polymer, because it is already widely used in 3D printing (4). Multiwalled carbon nanotubes were chosen as the conductive filler.

Several nanocomposites were prepared by dispersing various concentrations of multiwalled carbon nanotubes in ABS using two different methods: melt blending and solution casting.

In melt blending, shear deformation of the molten material is used to disperse the filler in the matrix. This deformation can orient the nanotubes. In contrast, solution casting is expected to randomly disperse the nanotubes. In addition, the degree of dispersion of the nanocomposite produced by the two methods need not be identical. The effect of the shear deformation was measured via the conductivity of the nanocomposites prepared by the two methods.

For both preparation methods, the addition of a small quantity of multiwalled carbon nanotube of 0.25% increased the conductivity by several orders of magnitude. Upon further addition of multiwalled carbon nanotube, the conductivity of the melt-blended samples increased more rapidly than that of the solution cast samples such that beyond approximately 1% of multiwalled carbon nanotube content, the conductivity of the melt-blended samples was almost two orders of magnitude higher than that of the solution cast samples.

In summary, an increase in conductivity of several orders of magnitude without significantly altering the rheology of the nanocomposite could be achieved. Therefore, it is expected that such materials can be used in conventional 3D printers, even in the case of narrow nozzles. While the conductivity obtained is still several orders of magnitude smaller than that of metals and hence not sufficient for applications such as circuits, the conductivity is sufficient for applications such as electromagnetic shielding and electrostatic discharge (4).

4.1.2 *Modular Supercapacitors*

The development of a multimaterial free-form 3D fabrication process can provide the possibility of building complete functional electronic devices (5). A design and manufacturing process has been described for electrochemical supercapacitors.

A combination of two 3D printing systems, i.e., a fused deposition modeling (FDM) printer and a paste extruder, were applied to fabricate these energy storage devices. During the manufacturing process of supercapacitor components, the FDM 3D printer was used to print the packaging frames, the conductive layers, and the electrode layers. Furthermore, the separator with electrolyte was deposited using the paste extruder. Several complete energy storage supercapacitors have been made and their electrochemical performances were assessed. The 3D manufacturing process developed was also evaluated in the study (5).

A 3D printing approach has been developed to fabricate quasi-solid-state asymmetric micro-supercapacitors to simultaneously realize the efficient patterning and ultrahigh areal energy density (6).

Typically, cathode, anode, and electrolyte inks with high viscosities and shear-thinning rheological behaviors are first prepared and 3D printed individually on the substrates. The 3D printed asymmetric micro-supercapacitor with interdigitated electrodes exhibits excellent structural integrity, a large areal mass loading of 3.1 $mg\,cm^{-2}$, and a wide electrochemical potential window of 1.6 V. Consequently, this 3D printed asymmetric micro-supercapacitor displays an ultrahigh areal capacitance of 207.9 $mF\,cm^{-2}$. More importantly, an areal energy density of 73.9 $\mu W\,cm^{-2}$ is obtained, which is superior to most reported interdigitated micro-supercapacitors. It is believed that the efficient 3D printing strategy can be used to construct various asymmetric micro-supercapacitors to promote the integration in on-chip energy storage systems (6).

4.1.3 *Active Electronic Materials*

The free-form generation of active electronics in unique architectures which transcend the planarity inherent in conventional microfabrication techniques has been an area of increasing scientific interest (7). Three-dimensional, large-scale integration can reduce the overall footprint and power consumption of electronics, and is usually accomplished via stacks of two-dimensional semiconductor wafers, in which interconnects between layers are achieved using wire-bonding or through-silicon vias.

Overcoming this 2D barrier has significant potential applications beyond improving the scalability in semiconductor integration technologies. For instance, the ability to seamlessly incorporate electronics into three-dimensional constructs could impart functionalities to biological and mechanical systems, such as advanced optical, computation or sensing capabilities.

For example, integration of electronics on otherwise passive structural medical instruments, such as catheters, gloves, and contact lenses, is critical for next-generation applications such as real-time monitoring of physiological conditions. Such integration has been previously demonstrated via meticulous transfer printing of prefabricated electronics and/or interfacing materials via dissolvable media such as silk on nonplanar surface topologies. An alternative approach is to attempt to interweave electronics in three dimensions from the bottom up. Yet, attaining seamless interweaving of electronics is challenging due to the inherent material incompatibilities and geometrical constraints of traditional microfabrication processing techniques (7).

Strategies directed at the seamless interweaving of three-dimensional active electronic devices have been reported (7).

These methods may include 3D printing of elastomeric matrices, organic polymers, solid and liquid metal leads, nanoparticle semiconductors, and/or transparent substrate layers. This method may also involve the identification of at least one material of an electrode, semiconductor, or polymer that possesses a desirable functionality and that exists in a printable format, and then patterning that material via direct dispensing from a computer-aided design (CAD)-designed construct onto a substrate.

In particular, the method consists of the steps of (7):

1. Providing an ink comprising semiconductor particles,
2. Extruding the ink comprising semiconductor particles to form at least one active electronic layer via 3D printing,
3. Providing a conductive ink, and
4. Extruding the conductive ink to form at least one conductive pattern via 3D printing, wherein at least one conductive pattern is adapted to allow an electric potential to be applied across the active electronic layer.

Conformal 3D printing, such as printing onto a curved surface like a contact lens, may also require scanning the topology of the surface of the substrate, and providing that information to the CAD system. The substrate can be a variety of desirable materials exhibiting flat or non-flat surface, such as biologics, glass, polyamides, or 3D printed substrates.

These semiconductors may provide a multitude of end uses, such as wearable displays and/or continuous on-eye glucose sensors. These devices may also include a range of functionalities, from quantum-dot-light-emitting diodes, MEMS devices, transistors, solar cells, piezoelectrics, batteries, fuel cells, and photodiodes.

For example, Poly-*N*,*N*′-bis-(3-methylphenyl)-*N*,*N*′-diphenylbenzidine (TPD) (American Dye Source, Quebec, Canada) was dissolved in chlorobenzene and dispensed on top of printed poly-(3,4-ethylenedioxythiophene):poly(styrene sulfonate). In contrast to spin coating, in which a significantly higher concentration of 1.5% is typically used as the starting solution, a device with appropriate diode characteristics was obtained when the concentration was reduced 10–18 fold (i.e., around 0.0825% to 0.15%) during direct ink 3D printing. Concentrations of 1.5% resulted in nonuniform films with bulk resistances that were too high for this particular application (2.44×10^9 Ω at 10 *V*).

Significantly diluted concentrations of poly-TPD of 0.015% yielded unstable device performances due to the discontinuous surface of the poly-TPD layer. It should be noted that although this particular combination had a workable range of between about

0.02% and about 0.20%, the exact maximum and minimum concentrations within which a given combination of materials and solvents will produce a continuous surface that is thin or thick enough to meet a particular application requirement will vary depending on the materials and solvents involved (7).

4.1.3.1 *Electrochemical Energy Storage*

In the field of electrochemical energy storage, 3D printing is of special interest because of its inherent advantages, including the freeform construction and controllable 3D structural prototyping (8).

The topics of 3D printed electrochemical energy storage devices have been summarized, which bridge advanced electrochemical energy storage and future additive manufacturing. Basic 3D printing systems and material considerations are described to provide a fundamental understanding of printing technologies for the fabrication of electrochemical energy storage devices.

The performance metrics of 3D printed electrochemical energy storage devices have been assessed and the related performance optimization strategies were discussed. Also, the recent advances of 3D printed electrochemical energy storage devices, including sandwich-type and in-plane architectures, were summarized. It can be expected that using 3D printing technology, the development of advanced electrochemical energy storage systems will be greatly promoted (8).

4.1.3.2 *Semiconducting Structures*

Poly(sulfone) (PSf) and poly(aniline) (PANI) composites were 3D printed using a solvent-cast direct write deposition. Traditional material extrusion techniques require the application of heat to melt the polymer during extrusion and printing. This type of thermal processing poses potential limitations for printing polymers that have high processing temperatures or thermally degrade.

Although PSf is a thermally stable polymer, it has an elevated glass transition temperature of 185–190°C and is highly viscous in the melt. On the other hand, PANI is a semiconducting polymer that thermally degrades before melting.

By using solvent-based inks, PSf and PSf/PANI composites were 3D printed at room temperature using direct write deposition. The PSf inks consisted of PSf dissolved in a mixture of dichloromethane and *N,N*-dimethylformamide (DMF). The dichloromethane evaporated quickly to harden the extruded filament, while the DMF evaporated slowly to allow for a smoother extruded filament and more consistent extrusion with well-bonded layers.

Best results were obtained with PSf concentrations of 35% to 40% with a dichloromethane:DMF volume ratio of 5:1. The optimized PSf/PANI inks consisted of 30% undoped PANI and 35% doped PANI, with 20% PSf solution in dichloroethane used as a binder.

By using capillary viscometry it was confirmed that the inks exhibited a pseudoplastic behavior, which is expected for polymer solutions and melts. It was shown that objects printed using the PSf/PANI ink are not conductive when undoped PANI was used, but became conductive when the PANI powder was first doped in 1 M H_2SO_4 before printing. A resistivity of 4.83 Ωm was achieved with an ink containing 35% doped PANI and 13% PSf (9).

4.1.4 *Piezoelectric Materials*

Piezoelectricity is a material property where electric charge accumulates in response to applied mechanical stress on the material (10). Materials that exhibit piezoelectricity include specific crystals, ceramics, and biological matter, such as deoxyribonucleic acid, proteins, and bone. The piezoelectric effect is a linear electromechanical interaction between the mechanical and electrical state in crystalline materials with no inversion symmetry.

Also, the piezoelectric effect is a reversible process. The direct piezoelectric effect is the internal generation of electrical charge resulting from an applied mechanical force, and the converse piezoelectric effect is the internal generation of a mechanical strain resulting from an applied electrical field. A valuable property of piezoelectric materials is an ability to convert compressive/tensile stresses to an electric charge, or vice versa.

Efficient piezoelectric nanoparticle-polymer composite materials can be optically printed into 3D microstructures using the method of digital projection printing (11).

Piezoelectric polymers were fabricated by incorporating barium titanate nanoparticles into photoliable polymer solutions such as poly(ethylene glycol) diacrylate and exposing to digital optical masks that could be dynamically altered to generate user-defined 3D microstructures.

Barium titanate nanoparticles were synthesized by hydrothermal methods (10):

Preparation 4–1: Precursors for the reaction include barium hydroxide monohydrate, titanium butoxide, and diethanolamine. First, 25 *m mol* of titanium butoxide was added to 10 *ml* of ethanol followed by the addition of 3.5 *ml* of ammonia. The titanium butoxide solution was then mixed with the barium hydroxide solution which contained 37.5 *m mol* of barium hydroxide in 12.5 *ml* distilled water. The diethanolamine (2.5 *ml*) was then added to the solution to help control the size of the nanoparticles.

The final solution was transferred to a Teflon-lined stainless-steel reactor and the reactor was kept in an oven at 200°C for 16 *h*. After the reaction, the reactor was cooled down to room temperature and the particles were cleaned 10 times with a vacuum filter using ethanol and distilled water. The final product was dried at 80°C for 24 *h*.

An example of a fabrication system for making 3D piezoelectric polymer structures based on light-assisted polymerization using UV light is shown in Figure 4.2.

The optical printing cells included cover glass slides coated with 100 *nm* of indium tin oxide deposited by magnetic sputtering. The electrodes were covered with approximately 1 *μm* of poly-(methyl methacrylate) to prevent shorting. A photoinitiator such as 2,2-dimethoxy-2-phenylacetophenone or 2,2-dimethoxy-1,2-di-(phenyl)ethanone (Irgacure 651) was added to the poly(ethylene glycol) diacrylate composites at a concentration of 1%. The poly-(ethylene glycol) diacrylate composite was then placed between the two electrodes using a 25 *μm* Kapton film spacer and the polymer could be polymerized using 365 *nm* light from a light-emitting diode (LED) or a hand-held UV lamp for film preparation. The power of the hand-held lamp was much lower than the LED.

The composite materials have to be activated in order to become piezoelectric.

In order to enhance the mechanical to electrical conversion efficiency of the composites, the barium titanate nanoparticles were chemically modified with acrylate surface groups, which formed

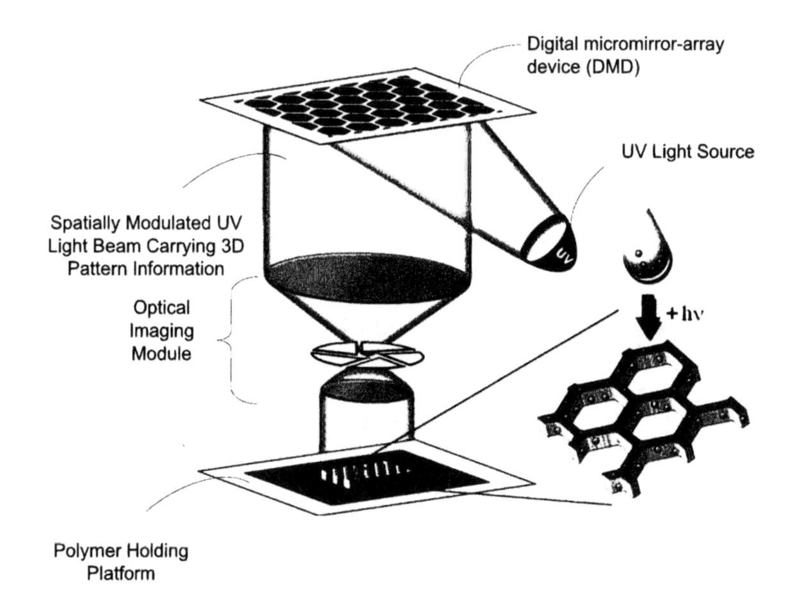

Figure 4.2 Fabrication system for making 3D piezoelectric polymer structures (10).

direct covalent linkages with the polymer matrix under light exposure. An example of such a modifier is a silane-based compound, i.e., 3-tri(methoxysilyl)propyl acrylate.

The composites with a 10% mass loading of the chemically modified barium titanate nanoparticles showed piezoelectric coefficients (d33) of around 40 pCN^{-1}, which were more than 10 times larger than composites synthesized with unmodified barium titanate nanoparticles and more than 2 times larger than composites that contained unmodified barium titanate nanoparticles and carbon nanotubes to boost mechanical stress transfer efficiencies (11).

In this way, methods for creating highly efficient piezoelectric polymer materials via nanointerfacial tuning were developed (11).

Exemplary applications of the here developed technology can include a wide range of devices, systems, and processes utilizing direct and/or converse piezoelectric effects, e.g., nonvolatile low voltage memory, loud speakers, acoustic imaging, energy harvesting, or electrical actuators (10).

For example, a 3D piezoelectric polymer can be used in pressure-sensing mouthguards for sports applications (10).

4.1.5 *Holographic Metasurface Antenna*

A 3D printed holographic metasurface antenna for beam focusing applications at 10 *GHz* within the X-band frequency regime has been described (12). The metasurface antenna is printed using a dual-material 3D printer leveraging a biodegradable conductive polymer material (Electrifi (13)) to print the conductive parts and PLA to print the dielectric substrate. The Electrifi conductive filament is a copper-polyester composite filament, which has a conductivity of $1.67 \times 10^4\ S\,m^{-1}$.

The entire metasurface antenna can be 3D printed at once. No additional techniques, such as metal-plating and laser etching, are required. It was demonstrated that by using the 3D printed conductive polymer metasurface antenna, high-fidelity beam focusing can be achieved within the Fresnel region of the antenna. It is also shown that the material conductivity for 3D printing has a substantial effect on the radiation characteristics of the metasurface antenna (12).

4.1.6 *Waveguide*

Terahertz spectroscopy grew significantly during the 1990s. Waveguides for the terahertz regime can be used for microscopy, terahertz quantum-cascade lasers, and imaging (14), but can also be useful in spectroscopy applications to replace complicated lens systems. A challenge in the design of a waveguide for the terahertz regime lies in the high material absorption for these frequencies (15).

In a hollow core antiresonant fiber, light is confined and guided in the air core by virtue of the resonant reflection provided by thin membranes that surround it and behave effectively as a Fabry-Perot cavity. Therefore, the transmission properties of such a waveguide depend on the thickness of these membranes and on the properties of the material from which they are made, specifically their refractive index.

A 3D printed waveguide has been developed that provides effective electromagnetic guidance in the *THz* regime (16). The waveguide is printed using low-cost poly(carbonate) (PC) and a conventional FDM printer. As the waveguides are fabricated using an FDM printer, the thickness of the membrane in the cladding depends on the nozzle size of the printer.

The cladding design used in the waveguides reported consists of semi-elliptical membranes, which have been shown to offer only a low loss at optical frequencies. Three different waveguides were fabricated. The parameters of these waveguides are shown in Table 4.1.

Table 4.1 Parameters of the waveguides (16).

Thickness/[mm]	Core size /[mm]	Ellipse length/[mm]
0.6	6	3.6
0.6	10	5
1.2	10	5

Light guidance in the hollow core is achieved through antiresonance, and it improves the energy effectively transported to the receiver compared to free-space propagation.

For the fabrication of the waveguides, PC was used, which has a refractive index in the terahertz range of 1.6 (17). A thickness of 0.6 mm fixes the position of the first four resonances at 0.2, 0.4, 0.6, 0.8 THz. A different nozzle would be required to print membranes of different thickness and shift the resonance peaks. Depending on the cladding design, the waveguide could be manufactured within 30 min at low cost.

4.1.7 *Fuel Cell*

A gas diffusion layer is an integral component of a polymer electrolyte membrane fuel cell stack, which plays a significant role in determining its performance, durability and the dynamic characteristics (18).

An ideal gas diffusion layer function to simultaneously transport three of the five essential elements, namely gas, water and heat, involved in the electrochemical reaction. In addition, it also transports the electron produced in the electrochemical reaction and serves as an armor to safeguard the membrane (Nafion), which is a delicate and most expensive component of the polymer electrolyte membrane fuel cell stack. However, the conventional carbon-based gas diffusion layer materials suffer from degradation issues during polymer electrolyte membrane fuel cell operation, and the predominant one is the electrochemical voltage oxidation.

The electrochemical degradation occurs due to the oxidation of the carbon present in the carbon paper to carbon dioxide especially at voltages greater than 0.207 *V* on a standard hydrogen electrode (SHE). The formation of carbon dioxide runs as (19)

$$C + 2H_2O \rightarrow CO_2^{+} 4H^{+} + 4e^{-} \quad . \tag{4.1}$$

Operating a polymer electrolyte membrane fuel cell stack at a low voltage (<0.207 *V*) is not practically possible since it can severely aggravate the operating efficiency and power density of the polymer electrolyte membrane fuel cell stack. The incorporation of a gas diffusion layer that is free from carbon can be a promising solution to circumvent these issues about the electrochemical oxidation. Also, the conventional gas diffusion layer manufacturing technique has a tedious and complicated process, which involves multiple stages. These multiple production stages also led to its high manufacturing costs and increased lead time.

Both these issues of gas diffusion layer durability and manufacturing costs have been assessed. The additive manufacturing method incorporating a selective laser sintering (SLS) technique provides a comprehensive solution to address both these issues. The concept of SLS is that the laser beam robotically scans the composite powder (base and conductive powder) at points in a space defined by a 3D model, fusing and subsequently binding the composite material together to create a solid-state structure.

Thus, SLS can be a favorable route to fabricate a carbon-free gas diffusion layer as well as to reduce its manufacturing costs and lead time. Finally, holistic characterization studies were performed to have a general insight on the characteristics of the proposed material. Valuable information is extrapolated from the characterization studies, which can assist in fine-tuning the material selection and SLS process parameter (18).

4.1.8 *Batteries*

4.1.8.1 *Microbattery Architectures*

Miniature power sources are needed for a variety of applications, including implantable medical devices, remote microsensors and transmitters, smart cards, and internet systems (20).

At the moment, rechargeable lithium-ion batteries, with the best performance on the subject of energy density and with a reasonably good power efficiency, dominate the consumer market.

Insufficient areal energy density from thin-film planar microbatteries has inspired a search for three-dimensional microbatteries using low-cost and efficient micro- and nanoscale materials and techniques. 3D printing technology enables the design of different shapes and high-surface-area structures, which no other manufacturing method can easily do.

A novel, quasi-solid rechargeable 3D microbattery has been presented, which is assembled on a 3D printed perforated polymer substrate with interconnected channels formed through XYZ planes. Simple and inexpensive electrophoretic-deposition routes can be applied for the fabrication of all the thin-film active-material layers of the microbattery. With the advantage of thin films, which conformally follow all the contours of the 3D substrate and are composed of nanosize electrode materials, such as modified lithium iron phosphate lithium titanate, and original polymer-in-ceramic electrolyte, these 3D microbatteries have a high reversible specific capacity and high pulse-power capability (20).

The applications of printed batteries have been discussed in a monograph (21). Applications are based on products that are mostly likely to benefit from the characteristics of printed batteries. In particular, the applications of microbatteries were detailed. These batteries have a footprint in the mm^2 range and they are useful for powering devices with a very small footprint. The applications of thin printed primary batteries were described. These batteries are thin, cheap, easy to fabricate, and useful for powering disposable types of electronics. Also, applications of high-performance rechargeable flexible batteries were explained. These batteries have electrochemical performance comparable to conventional batteries. The structured design helps to improve the energy density and rate performance of the current generation of batteries. Also, the power electronics required for charging printed batteries with harvesters were explained (21).

The 3D printing technique allows functional inks to be precisely patterned in filamentary form (22).

To print high-aspect-ratio electrode architectures, the composition and rheology of each ink must be optimized to ensure a reliable flow through the nozzles and to promote adhesion between the printed features.

Concentrated anode and cathode inks have been prepared by suspending $Li_4Ti_5O_{12}$ and $LiFePO_4$ in a solution composed of deionized water, ethylene glycol, glycerol, and cellulose-based viscosifiers.

After printing, the dried microelectrode arrays are heated to 600°C in inert gas to remove the organic additives and to promote nanoparticle sintering. Thermogravimetry reveals that the organic species are largely removed at 300°C.

At higher temperatures, the particles undergo sintering, which leads to neck formation at particle-particle contacts. The annealed structures remain highly porous, which is desirable for electrolyte penetration (22).

Rechargeable lithium-ion batteries with a higher capacity in customized geometries have been developed (23).

The design, fabrication, and electrochemical performance of fully 3D printed lithium-ion batteries composed of thick semisolid electrodes that exhibit high areal capacity were reported. Semisolid cathode and anode inks, as well as UV curable packaging and separator inks for direct writing of lithium-ion batteries in arbitrary geometries, have been created.

These fully 3D printed and packaged lithium-ion batteries, which are encased between two glassy carbon current collectors, deliver an areal capacity of 4.45 $mA\,h\,cm^{-2}$ at a current density of 0.14 $mA\,cm^{-2}$, which is equivalent to 17.3 $A\,h\,L^{-2}$. The ability to produce such high-performance batteries in customized form opens new avenues for integrating batteries directly within 3D printed objects (23).

4.1.8.2 3D Printing Electrolytes

Solid-state batteries have many advantages in terms of safety and stability (24). However, the solid electrolytes upon which these batteries are based typically lead to a high cell resistance. Both components of the resistance (interfacial, due to poor contact with electrolytes, and bulk, due to a thick electrolyte) are a result of the rudimentary manufacturing capabilities that exist for solid-state electrolytes.

In general, solid electrolytes are studied as flat pellets with planar interfaces, which minimize interfacial contact area. Multiple ink formulations have been developed that enable 3D printing of unique solid electrolyte microstructures with varying properties (24).

These inks are used to 3D print a variety of patterns, which are then sintered to reveal thin, nonplanar, intricate architectures composed only of $Li_7La_3Zr_2O_{12}$ solid electrolyte. Using these 3D printing ink formulations to further study and optimize electrolyte structure could lead to solid-state batteries with dramatically lower full cell resistance and higher energy and power density.

In addition, the reported ink compositions could be used as a model recipe for other solid electrolyte or ceramic inks, perhaps enabling 3D printing in related fields (24).

4.1.8.3 *Porous Battery Electrodes*

A highly porous 2D nanomaterial, holey graphene oxide, could be synthesized directly from holey graphene powder and employed to create an aqueous 3D printable ink without the use of additives or binders (25).

Stable dispersions of hydrophilic holey graphene oxide sheets in water (100 $mg\,ml^{-1}$) could be readily achieved. The shear-thinning behavior of the aqueous holey graphene oxide ink enables the extrusion-based printing of fine filaments into complex 3D architectures, such as stacked mesh structures, on arbitrary substrates. The free-standing 3D printed holey graphene oxide meshes exhibit trimodal porosity: nanoscale, i.e., 4 nm to 25 nm through-holes on holey graphene oxide sheets, microscale (tens of micrometer sized pores introduced by lyophilization), and macroscale ($> 500\ \mu m$ square pores of the mesh design), which are advantageous for high-performance energy storage devices that rely on interfacial reactions to promote full active-site utilization.

The additive-free architectures have been demonstrated as the first 3D printed lithium oxygen ($Li{-}O_2$) cathodes and characterized alongside 3D printed GO-based materials without nanoporosity as well as nanoporous 2D vacuum filtrated films.

The results of the study indicated a synergistic effect between 2D nanomaterials, hierarchical porosity, and overall structural design, as well as the promise of a free-form generation of high-energy-density battery systems (25).

Porous Carbon Anode. 3D porous carbon structures, fabricated via a 3D printing technique, could be utilized as the anode materials for microbial fuel cells (26). The intrinsic biocompatibility of 3D printed carbon anodes, together with the open porous structures, greatly enhanced the metabolic activities of microorganisms. The secondary 3D roughness generated from carbon formation functioned as an ideal support for microbial growth, which further increased the surface area of anodes as well.

All these factors resulted in exclusive electrochemical performances of microbial fuel cells for enhanced power generation and scaling up application. Through carefully tuning the carbonization processes, a multiscale 3D porous carbon structure could be achieved for bacterial growth and mass transfer, leading to the highest maximum output voltage, open circuit potential and a power density for a 300 μm porosity ($453.4 \pm 6.5\ mV$, $1256 \pm 69.9\ mV$ and $233.5 \pm 11.6\ mW\,m^{-2}$, respectively). Such performance is superior to that of carbon cloth anode and carbon fiber brush anode under the same condition (26).

4.1.8.4 Printed Batteries

The printing of solid polymer and gel electrolytes is still a bottleneck in the all-printed batteries (27).

The electrochemical performance of conventional bulk batteries and printed batteries with polymer electrolytes was compared. Conventional batteries are fabricated by superposition of positive and negative electrodes coated on current collectors with an intermediate layer of a separator or polymer electrolyte.

Screen printing has been applied mainly to the fabrication of positive and negative electrodes for lithium-ion batteries and supercapacitors. The rheology requirements of screen printing inks are similar to those of a standard battery. Spray printing is attractive for sequentially printing multiple inks that share a common solvent, one on top of the other. The main types of inkjet printing processes, which are related to direct printing methods, are continuous and drop-on-demand (27).

4.1.8.5 *Hybrid Electrolytes*

Ceramic electrolytes have seen major advances in conductivity, with the sulfides, in particular, achieving conductivities comparable to liquids (28).

However, ceramics exhibit typical mechanical properties that make maintaining fracture-free membranes that retain contact with the electrode a challenge. In short, the transition to solid electrolytes introduces the problem of mechanical properties and each class of solid electrolyte presents its own, largely complementary advantages and disadvantages.

In recognition of this complementarity, efforts have been made to combine ceramic and polymer electrolytes to form composite electrolytes (29).

Hybrid solid electrolytes composed of 3D ordered bicontinuous conducting ceramic and insulating polymer microchannels have been described (29).

The ceramic channels provide continuous, uninterrupted pathways, maintaining high ionic conductivity between the electrodes, while the polymer channels permit an improvement of the mechanical properties from that of the ceramic alone, in particular mitigation of the ceramic brittleness.

The conductivity of a ceramic electrolyte is usually limited by resistance at the grain boundaries, necessitating dense ceramics. The conductivity of the 3D ordered hybrid is reduced by only the volume fraction occupied by the ceramic. This demonstrates that the ceramic channels can be sintered to high density similar to a dense ceramic disk. The hybrid electrolytes are demonstrated using the ceramic lithium-ion conductor $Li_{14}\,4Al_{4}\,44Ge_{16}(PO_{4})_{3}$.

The $Li_{14}\,4Al_{4}\,44Ge_{16}(PO_{4})_{3}$ powder was prepared using the Pechini method (30,31), which is a sol-gel process involving a chelating polymer that is effective in distributing the ions on an atomic scale. Certain α-hydroxycarboxylic acids, such as citric, lactic and glycolic acids, can form polybasic acid chelates with the metals. The chelates can undergo a polyesterification when heated with a polyhydroxy alcohol.

Structured Li_{14} $4Al_4$ $4Ge_{16}(PO_4)_3$ 3D scaffolds with empty channels were prepared by negative replication of a 3D printed polymer template. Filling the empty channels with non-conducting poly(propylene) or an epoxy resin creates the structured hybrid electrolytes with 3D bicontinuous ceramic and polymer microchannels. The printed templating method permits a precise control of the ceramic-to-polymer ratio and the microarchitecture. This can be demonstrated by the formation of cubic, gyroidal, diamond and spinodal structures. The parameters that characterize the 3D printed templates and structured $Li_{14}4Al_44Ge_{16}(PO_4)_3$ (LAGP) scaffolds are shown in Table 4.2.

The electrical and mechanical properties depend on the microarchitecture with the gyroid filled with epoxy showing the best combination of conductivity and mechanical properties. An ionic conductivity of $1.6\times10-4\ S\ cm^{-1}$ at room temperature was obtained, reduced from the conductivity of a sintered $Li_{14}4Al_44Ge_{16}(PO_4)_3$ pellet only by the volume fraction occupied by the ceramic. The mechanical properties of the gyroid $Li_{14}4Al_44Ge_{16}(PO_4)_3$-epoxy electrolyte demonstrate up to 28% higher compressive failure strain and up to five times the flexural failure strain of a $Li_{14}4Al_44Ge_{16}(PO_4)_3$ pellet before rupture.

These findings demonstrate that ordered ceramic and polymer hybrid electrolytes can have superior mechanical properties without significantly compromising ionic conductivity, which addresses one of the key challenges for all-solid-state batteries (29).

4.2 Magnetic Uses

4.2.1 *Polymer-Based Permanent Magnets*

The production of permanent magnets is a complex process that implies very specific machinery able to perform very specific actions (32). Even though this problem has been approached by extrusion-based three-dimensional printing techniques, the resulting parts either have not been fully characterized or present low-resolution outputs.

Table 4.2 Parameters of the 3D printed templates and structured LAGP scaffolds (29).

Material	Property	Cube	Gyroid	Diamond	Bijet
3D printed template	Volume shrinkage during printing /[%]	0	2	6	2
Template LAGP filled	Volume expansion during filling /[%]	16	16	13	24
Structured LAGP scaffold	Volume shrinkage during sintering /[%]	44	40	35	40
Structured LAGP scaffold	Overall volume shrinkage /[%]	34	32	31	28
Template design	Solid volume fraction /[%]	15.6	15.6	15.4	15.6
Structured LAGP scaffold	Solid volume fraction /[%]	71.4	72.6	72.5	78.2
Template design	Average empty channel diameter /[μm]	66.1	66.4	67.8	69.0
Structured LAGP scaffold	Average LAGP thickness /[μm]	44.2	44.0	42.0	59.5
Template design	Average wall thickness /[μm]	15.1	20.4	19.9	22.5
Structured LAGP scaffold	Average empty channel diameter /[μm]	20.4	20.8	18.4	22.3
Structured LAGP scaffold	Weight /[mg]	5.54	6.60	5.90	5.94
Structured LAGP scaffold	Volume /[m^3]	2.27	2.78	2.59	2.50
Structured LAGP scaffold	Density /[$mg\,mm^{-3}$]	2.44	2.37	2.28	2.38

A stereolithography three-dimensional printing technique was used to address this problem. It could be demonstrated that it is possible to develop high-resolution polymer-based permanent magnets.

Some of the basic materials used in the study are shown in Table 4.3 and in Figure 4.3.

Table 4.3 Basic materials (32).

Compound
Iron oxide (Fe_2O_3) nanoparticles of 50–100 *nm* size
Iron oxide (Fe_2O_3) nanoparticles of 25 *nm* size (US Research Nanomaterials Inc.)
Bisphenol A ethoxylate dimethacrylate
Poly(ethylene glycol) diacrylate
Phenylbis(2,4,6-trimethylbenzoyl)phosphine oxide (Irgacure 819)
Triarylsulfonium hexafluorophosphate salts mixed 50% in propylene carbonate
Sudan I (Sigma-Aldrich)

Bisphenol A ethoxylate dimethacrylate is an organic polymerizable methacrylate group linked to bisphenol by oligo-groups of different length, which determines the glass transition temperature of the thermoset networks after exposure to UV light. Poly(ethylene glycol) diacrylate is a hydrophilic hydrogel in which the crosslinker also acts as a crosslinked chain (33).

Photochemical initiators have been used to crosslink the unsaturated monomers through the homolysis process Irgacure 819 was added, separately, in both bisphenol A ethoxylate dimethacrylate and poly(ethylene glycol) diacrylate at 1% and the mixture was left under magnetic stirring for 24 *h*. To improve 3D printing resolution, 0.1% of Sudan I was added into the mixture, as its light-sensitive azobenzene moieties block UV light from scattering (34), therefore producing better 3D printing resolution. The same process was repeated replacing Irgacure 819 with triarylsulfonium hexafluorophosphate salts. The polymers containing Irgacure 819 presented a yellowish color whereas those containing TH salts appeared transparent. When adding Sudan I to the mixtures both of them became orange.

UV-light photo-sensible magnetic composites were developed by

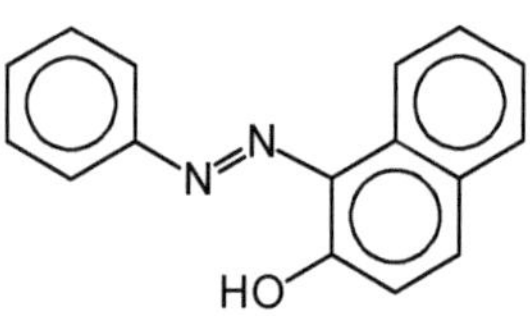

Sudan I

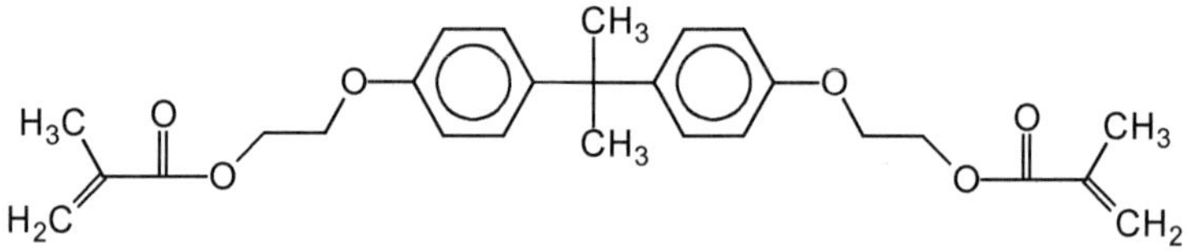

Bisphenol A ethoxylate dimethacrylate

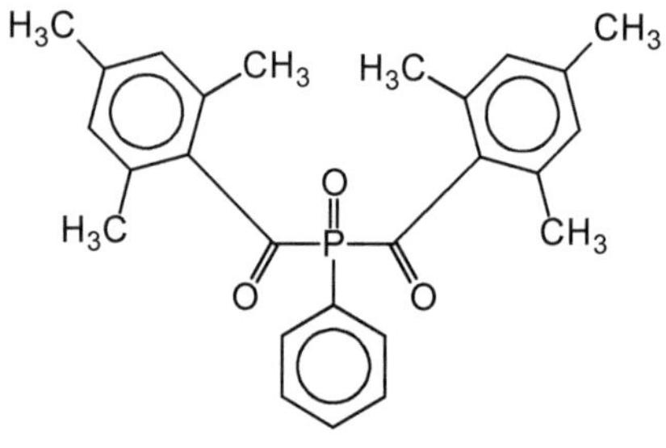

Phenylbis(2,4,6-trimethylbenzoyl)phosphine oxide

Figure 4.3 Basic materials.

embedding Fe_3O_4 and Fe nanoparticles within the different polymers at different amounts of 25%, 50%, and 75%.

In a mechanical mixer, the composites were mixed for 3 *min* at 1500 *rpm*, and defoamed for 2 *min* at 1200 *rpm*. The composites that contained Fe_3O_4 nanoparticles exhibited a black color and the composites containing Fe nanoparticles exhibited a grey color.

All the samples were developed using a screen printing approach. When the polymer is cured under UV light it expands, leading to slight variations of the thickness that need to be accounted for when developing CAD files. Thus, obtaining a good balance between 3D printing resolution and physical properties is vital during the fabrication process. Whereas higher concentrations of nanoparticles will produce a better magnetic output, this same parameter will lead to poor 3D printing resolution due to an increase of light scattering and agglomeration of nanoparticles (32).

Both size and thickness of the 3D printed samples generate different magnetic fields. The larger the 3D printed sample is, the higher magnetic field it generates. A 10 *mm* side length sample of 1 *mm* thickness generated a magnetic field lower than that generated by the sample of the same side length and twice its thickness. This is associated with the higher content of magnetic material. On the other hand, it was observed that reducing the side length to 5 *mm* and increasing the thickness to 2.5 *mm*, resulted in the magnetic field being reduced by a factor of 10. This was associated with the reduction of magnetic material due to the decrease of the sample's volume. In this way, the produced magnetic field was observed to be strongly dependent on the shape and size of the 3D printed samples (32).

4.2.2 *Bonded Magnets*

A method has been reported for the fabrication of Nd–Fe–B bonded magnets of a complex shape using 3D printing (35).

A 3D printable epoxy-based ink was successfully formulated for direct-write additive manufacturing with anisotropic Magnequench (MQA) Nd–Fe–B magnet particles that can be deposited at room temperature. The new feedstocks contain up to 40% by volume MQA anisotropic Nd–Fe–B magnet particles, and they have been shown to remain uniformly dispersed in the thermoset matrix throughout the deposition process.

Ring, bar, and horseshoe-type 3D magnet structures were printed and cured in air at 100°C without degrading the magnetic properties. So, the study provided a new pathway for fabricating Nd–Fe–B bonded magnets with complex geometry at low temperature, and new opportunities were assessed for fabricating multifunctional hybrid structures and devices (35).

4.2.3 *Strontium Ferrite*

In a study, the orientation control of randomly shaped, anisotropic hard magnetic ferrite particles was demonstrated for material jetting-based additive manufacturing processes using a developed particle alignment configuration (36).

Strontium ferrite and PR48 photosensitive resin were used as the base materials.

An automated experimental setup with two neodymium permanent cube magnets capable of generating a dipolar magnetic field was built to align magnetic particles in the resin.

The alignment of the particles was characterized for directionality using images obtained through real-time optical microscopy. The orientation of magnetic particles was observed to be dependent on the distance of separation between the cube magnets and the magnetization time. Furthermore, X-ray diffraction was used to indicate the c-axis alignment of the hexagonal strontium ferrite particles in the cured specimens. The influence of process parameters on particle orientation was evaluated, using a full factorial experiment analysis (36).

4.2.4 *Soft-Magnetic Composite*

A 3D printed polymer-metal soft-magnetic composite was developed and characterized for its material, structural, and functional properties (37). The composition contains ABS as the polymer matrix, with up to 40% by volume of stainless-steel micropowder as the filler.

The composites were prepared by heated kneading of the ABS pellets with the steel micropowder in a Brabender W50-EHT kneader

(Brabender GmBH, Duisburg, Germany) for 60 *min* at 180°C. Then, the composite feedstocks were dried in an oven for 24 *h* at 130°C to minimize the moisture. Moisture was found to adversely affect the extrusion process due to the formation of voids.

The feedstocks were extruded into filaments using a Noztek Pro single barrel extruder (Noztek, Shoreham, England) at temperatures between 190°C and 210°C. Pure ABS filaments were also extruded at 190°C for comparison. The target filament diameter of 1.75 ± 10 *mm* was achieved by adjusting the spooling speed of the extruder.

The composites were rheologically analyzed and 3D printed into tensile and flexural test specimens using a commercial desktop 3D printer. The mechanical characterization revealed a linearly decreasing trend of the ultimate tensile strength and a sharp decrease in Young's modulus with increasing filler content.

A four-point bending analysis showed a decrease of up to 70% in the flexural strength of the composite and up to a two-factor increase in the secant modulus of elasticity. Magnetic hysteresis characterization revealed retentivities of up to 15.6 *mT* and coercive forces of up to 4.31 $kA\ m^{-1}$ at an applied magnetic field of 485 $kA\ m^{-1}$.

The particle size distribution of a steel micropowder with a d_{50} value of 5.88 μm is shown in Figure 4.4.

The investigated composite is interesting as a material for the additive manufacturing of passive magnetic sensors and/or actuators (37).

4.2.5 *Discontinuous Fiber Composites by 3D Magnetic Printing*

Discontinuous fiber composites are a class of materials that are strong, lightweight and have remarkable fracture toughness (38). These advantages can partially explain the abundance and variety of discontinuous fiber composites that have evolved in the natural world.

Natural composites utilize reinforcing particles exquisitely organized into complex architectures to achieve superior mechanical properties, including the shells of abalones (39), the dactyl clubs of peacock mantis shrimp (40, 41), and the cortical bones of mammals (42). Such ordered, yet heterogeneous, reinforcement architectures are frequently linked to superior mechanical properties.

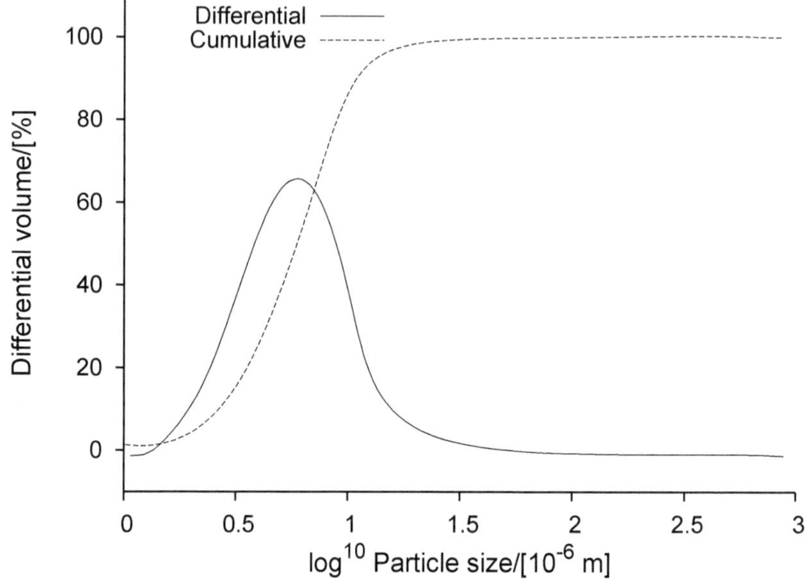

Figure 4.4 Particle size distribution of a steel micropowder (37).

Natural manufacturing processes are complex, however, colloidal assembly techniques and advances in additive manufacturing can be harnessed to grow synthetic composites with similarly complex architectures.

An additive manufacturing approach has been presented that combines real-time colloidal assembly with existing additive manufacturing technologies to create highly programmable discontinuous fiber composites (38). This technology is addressed as *3D magnetic printing*. It can enable the recreation of complex bioinspired reinforcement architectures that deliver an enhanced material performance in comparison to monolithic structures.

3D magnetic printing is capable of printing dense ceramic/polymer composites in which the direction of the ceramic-reinforcing particles can be finely tuned within each individual voxel of the printed material. To enable the orientation of the ceramic-reinforcing particles, magnetic labeling techniques are employed to coat traditionally nonmagnetic-reinforcing materials, such as high-strength alumina, silica and calcium phosphate, with nominal amounts of iron oxide nanoparticles (43,44). 3D magnetic printing can produce a wide variety of composite microarchitectures with clearly printed features down to 90 μm size which exhibit unique reinforcement.

Elegant reinforcement architectures of ceramic microparticles could be produced that exhibit feature sizes of 90 μm. These architectures enable the production of composite materials that exhibit increased stiffness, strength and hardness properties.

This method is robust, low cost, scalable, sustainable and can enable a new class of strong, lightweight composite prototypes with programmable properties (38).

References

1. G. Postiglione, G. Natale, G. Griffini, M. Levi, and S. Turri, *Composites Part A: Applied Science and Manufacturing*, Vol. 76, p. 110, 2015.
2. P.-C. Ma, N.A. Siddiqui, G. Marom, and J.-K. Kim, *Composites Part A: Applied Science and Manufacturing*, Vol. 41, p. 1345, 2010.
3. M.A. Cruz, S. Ye, M.J. Kim, C. Reyes, F. Yang, P.F. Flowers, and B.J. Wiley, *Particle & Particle Systems Characterization*, Vol. 35, p. 1700385, 2018.
4. T. Kobayashi, A. Khosla, M. Sugimoto, H. Furukawa, and S.K. Sukumaran, Electrically conductive polymer nanocomposites for 3D printing, in *ESC Meeting Abstracts*, number 3, pp. 192–192. The Electrochemical Society, 2018.
5. A. Tanwilaisiri, Y. Xu, R. Zhang, D. Harrison, J. Fyson, and M. Areir, *Journal of Energy Storage*, Vol. 16, p. 1, 2018.
6. K. Shen, J. Ding, and S. Yang, *Advanced Energy Materials*, Vol. 0, p. 1800408, 2018.
7. M.C. McAlpine and Y.L. Kong, 3D printed active electronic materials and devices, US Patent 9 887 356, assigned to The Trustees of Princeton University (Princeton, NJ), February 6, 2018.
8. X. Tian, J. Jin, S. Yuan, C.K. Chua, S.B. Tor, and K. Zhou, *Advanced Energy Materials*, Vol. 7, p. 1700127, 2017.
9. Z. Miao, J. Seo, and M.A. Hickner, *Polymer*, Vol. 152, p. 18, 2018.
10. D.J. Sirbuly, S. Chen, K. Kim, and W. Zhu, 3D piezoelectric polymer materials and devices, US Patent Application 20 160 322 560, assigned to University of California, November 3, 2016.
11. K. Kim, W. Zhu, X. Qu, C. Aaronson, W.R. McCall, S. Chen, and D.J. Sirbuly, *ACS Nano*, Vol. 8, p. 9799, 2014.
12. O. Yurduseven, S. Ye, T. Fromenteze, D.L. Marks, B.J. Wiley, and D.R. Smith, *Applied Physics*, pp. 1–4, 2018. preprint arXiv:1806.00394.
13. Multi3D, Electrifi conductive 3D printing filament, electronic: https://www.multi3dllc.com/about-electrifi/, 2016.

14. D.W. Vogt and R. Leonhardt, *Journal of Infrared, Millimeter, and Terahertz Waves*, Vol. 37, p. 1086, 2016.
15. A. Argyros, *ISRN Optics*, Vol. 2013, 2013.
16. L.D. van Putten, J. Gorecki, E.N. Fokoua, V. Apostolopoulos, and F. Poletti, *Appl. Opt.*, Vol. 57, p. 3953, May 2018.
17. M. Naftaly and R.E. Miles, *Proceedings of the IEEE*, Vol. 95, p. 1658, 2007.
18. A. Jayakumar, *A novel technique to manufacture carbon-free gas diffusion layer for polymer electrolyte membrane fuel cell application by a selective laser sintering (3D Printing)*. PhD thesis, Auckland University of Technology, 2018.
19. A. Jayakumar, *MATEC Web of Conferences*, Vol. 172, p. 04005, 2018.
20. E. Cohen, S. Menkin, M. Lifshits, Y. Kamir, A. Gladkich, G. Kosa, and D. Golodnitsky, *Electrochimica Acta*, Vol. 265, p. 690, 2018.
21. A.M. Gaikwad, A.E. Ostfeld, and A.C. Arias, Applications of printed batteries in *Printed Batteries*, chapter 6, pp. 144–184. Wiley-Blackwell, 2018.
22. K. Sun, T.-S. Wei, B.Y. Ahn, J.Y. Seo, S.J. Dillon, and J.A. Lewis, *Advanced Materials*, Vol. 25, p. 4539, June 2013.
23. T.-S. Wei, B.Y. Ahn, J. Grotto, and J.A. Lewis, *Advanced Materials*, Vol. 30, p. 1703027, 2018.
24. D.W. McOwen, S. Xu, Y. Gong, Y. Wen, G.L. Godbey, J.E. Gritton, T.R. Hamann, J. Dai, G.T. Hitz, L. Hu, and E.D. Wachsman, *Advanced Materials*, Vol. in press, p. 1707132, 2018.
25. S.D. Lacey, D.J. Kirsch, Y. Li, J.T. Morgenstern, B.C. Zarket, Y. Yao, J. Dai, L.Q. Garcia, B. Liu, T. Gao, S. Xu, S.R. Raghavan, J.W. Connell, Y. Lin, and L. Hu, *Advanced Materials*, Vol. 30, p. 1705651, 2018.
26. B. Bian, D. Shi, X. Cai, M. Hu, Q. Guo, C. Zhang, Q. Wang, A.X. Sun, and J. Yang, *Nano Energy*, Vol. 44, p. 174, 2018.
27. E. Strauss, S. Menkin, and D. Golodnitsky, Polymer electrolytes for printed batteries in S. Lanceros-Méndez and C.M. Costa, eds., *Printed Batteries*, chapter 4, pp. 80–111. Wiley-Blackwell, 2018.
28. N. Kamaya, K. Homma, Y. Yamakawa, M. Hirayama, R. Kanno, M. Yonemura, T. Kamiyama, Y. Kato, S. Hama, K. Kawamoto, and A. Mitsui, *Nature Materials*, Vol. 10, p. 682, July 2011.
29. S. Zekoll, C. Marriner-Edwards, A.O. Hekselman, J. Kasemchainan, C. Kuss, D.E. Armstrong, D. Cai, R.J. Wallace, F.H. Richter, J.H. Thijssen, et al., *Energy & Environmental Science*, Vol. 11, p. 185, 2018.
30. M.P. Pechini, Method of preparing lead and alkaline earth titanates and niobates and coating method using the same to form a capacitor, US Patent 3 330 697, assigned to Sprague Electric Co., July 11, 1967.
31. V. Petrykin and M. Kakihana, Chemistry and applcations of polymeric gel applcations in S. Sakka and H. Kozuka, eds., *Handbook of Sol-Gel*

Science and Technology, chapter 4, p. 77. Kluwer Academic Publishers, New York, 2005.
32. R. Domingo-Roca, J.C. Jackson, and J.F.C. Windmill, *Materials & Design*, Vol. 153, p. 120, 2018.
33. J.A. Beamish, J. Zhu, K. Kottke-Marchant, and R.E. Marchant, *Journal of Biomedical Materials Research - Part A*, Vol. 92, p. 441, 2010.
34. R. Domingo-Roca, B. Tiller, J.C. Jackson, and J.F.C. Windmill, *Sensors and Actuators, A: Physical*, Vol. 271, p. 1, 2018.
35. B.G. Compton, J.W. Kemp, T.V. Novikov, R.C. Pack, C.I. Nlebedim, C.E. Duty, O. Rios, and M.P. Paranthaman, *Materials and Manufacturing Processes*, Vol. 33, p. 109, 2018.
36. B. Nagarajan, A.F.E. Aguilera, M. Wiechmann, A.J. Qureshi, and P. Mertiny, *Additive Manufacturing*, Vol. 22, p. 528, 2018.
37. B. Khatri, K. Lappe, D. Noetzel, K. Pursche, and T. Hanemann, *Materials*, Vol. 11, p. 189, 2018.
38. J.J. Martin, B.E. Fiore, and R.M. Erb, *Nature Communications*, Vol. 6, p. 8641, 2015.
39. A.P. Jackson, J.F.V. Vincent, and R.M. Turner, *Proceedings of the Royal Society of London B: Biological Sciences*, Vol. 234, p. 415, 1988.
40. J.C. Weaver, G.W. Milliron, A. Miserez, K. Evans-Lutterodt, S. Herrera, I. Gallana, W.J. Mershon, B. Swanson, P. Zavattieri, E. DiMasi, and D. Kisailus, *Science*, Vol. 336, p. 1275, 2012.
41. L.K. Grunenfelder, N. Suksangpanya, C. Salinas, G. Milliron, N. Yaraghi, S. Herrera, K. Evans-Lutterodt, S.R. Nutt, P. Zavattieri, and D. Kisailus, *Acta Biomaterialia*, Vol. 10, p. 3997, 2014.
42. S. Weiner and H.D. Wagner, *Annual Review of Materials Science*, Vol. 28, p. 271, 1998.
43. R.M. Erb, R. Libanori, N. Rothfuchs, and A.R. Studart, *Science*, Vol. 335, p. 199, 2012.
44. R.M. Erb, K.H. Cherenack, R.E. Stahel, R. Libanori, T. Kinkeldei, N. Münzenrieder, G. Tröster, and A.R. Studart, *ACS Applied Materials & Interfaces*, Vol. 4, p. 2860, 2012.

5

Medical Applications

Medical applications of inkjet technology, e.g., drug delivery, have been detailed in a separate chapter. Here we focus on medical applications of 3D printing.

The potential impact on biotechnology of 3D printing has been reviewed (1).

Also, the uses and applications of 3D printing for tissue engineering have been reviewed (2). Biodegradable materials can also be printed that have proven to be ideal for bone tissue engineering, sometimes even with site-specific growth factor and drug delivery abilities.

Furthemore, the scaffold design and stem cells for tooth regeneration have been reviewed (3). Besides electrospinning, supercritical fluid-gassing and self-assembling, common rapid prototyping technology can also be used for the fabrication of a tooth scaffold.

5.1 Basic Procedures

The generation of graspable three-dimensional objects applied for surgical planning, prosthetics and related applications using 3D printing or rapid prototyping have been reviewed (4).

Source data acquired with any imaging modality can be typically visualized in two dimensions. With post-processing tools and algorithms, it is possible to produce multiplanar reformations and three-dimensional views of the anatomy.

The process chain involved from image acquisition to the production of a 3D rapid prototype model consists of the following three steps:

1. Image acquisition,
2. Image post-processing, and
3. 3D printing.

5.1.1 *Image Acquisition*

Image acquisition is an important step in the generation of 3D objects as the quality of the object depends on the quality of these data. Clinical image acquisition can be done at a ultrahigh spatial resolution of 400 μm –600 μm, with a good quality contrast.

A slice thickness of less than 1 mm and isotropic voxels are important parameters for minimizing the partial volume effect during post-processing (5).

Although multidetector computed tomography (MDCT) and MRI are equally good imaging modalities for data acquisition, MDCT is widely applied for rapid prototyping because image post-processing is less complex for MDCT data.

Cone beam computed tomography, positron emission tomography, single-photon emission computed tomography and ultrasonography are other noninvasive imaging modalities that can be used for data acquisition. Finally, in this step, the acquired data are saved in the common digital imaging and communications in medicine (DICOM) format.

5.1.1.1 *Image Post-Processing*

Special high-performance workstations equipped with post-processing tools are used for processing the DICOM images that were generated during acquisition. The common 3D post-processing tools include segmentation tools often based on simple region growing, as well as visualization tools such as surface volume rendering, maximal/minimal intensity projection and multiplanar reformation.

A wide area of application is in the surgical field, e.g., vascular surgery, orthopedic surgery, pediatric surgery, where these tools are routinely used in clinical practice for planning and executing surgery (6,7).

Additionally, advanced post-processing algorithms have been proposed, e.g., for low-resolution or nonenhanced images (8).

The contours of a segmented region of interest can be computationally transformed into a 3D triangle mesh (9), i.e., the shape of a part is approximated using triangular facets. Obviously, tiny triangular facets produce a smoother surface but increase the size of the data. The mesh data may be further processed using CAD (computer-aided design) software. This may include automatic mesh optimization or manual modifications of the geometry. Finally, the data is sent to the 3D printing machine for production, where the STL (surface tessellation language) file format is commonly used.

5.1.2 3D Printing

The 3D printing methodology uses three-dimensional CAD data sets for producing a 3D haptic physical model. It is also referred to as rapid prototyping, solid free form, computer automated or layered manufacturing depending on the kind of production method used. The principle of rapid prototyping is to use 3D computer models for the reconstruction of a 3D physical model by the addition of material layers [11]. With additive fabrication, the machine reads in data from a CAD drawing and lays down successive layers of liquid, powder, or the sheet material, and in this way builds up the model from a series of cross sections

These layers, which correspond to the virtual cross section from the CAD model joined together, create the final shape. The primary advantage of additive fabrication is its ability to create almost any complex shape or geometric feature. The word rapid has to be taken rather relatively: construction of a model with contemporary methods can take from several hours to days, whereas additive systems for rapid prototyping can typically produce models in a few hours. The eventual construction time depends on the specific method used, as well as the size and complexity of the model.

Rapid prototyping includes a number of established manufacturing techniques and a multitude of experimental technologies either in development or used by small groups of individuals. Each technique has its own limitations and applications in producing prototype models. An overview of the established rapid prototyping techniques used in the medical applications is given in Table 5.1.

Table 5.1 Rapid prototyping techniques used in medical fields (4).

Method	Advantage	Disadvantage
Stereolithography (SLA)	Large part size	Moderate strength
Selective laser sintering (SLS)	Large part size, variety of materials, good strength	High cost, powdery surface
Fused deposition modeling (FDM)	Low cost, good strength	Low speed
Laminated object manufacturing (LOM)	Low cost, large part size	Limited materials
Inkjet printing techniques	Low cost, high speed, multimaterial capability	Moderate strength

Stereolithography uses photopolymers that can be cured by UV laser. Selective laser sintering (SLS) is based on small particles of thermoplastic, metal, ceramic or glass powders that are fused by high-power laser systems. The materials include polymers such as nylon, glass-filled nylon or poly(styrene), or metals such as steel, stainless-steel alloys, bronze alloys or titanium. Fused deposition modeling (FDM) works by extruding small beads of fused thermoplastic materials or eutectic metals that immediately bond to the layer below. Laminated object manufacturing uses layers of paper or plastic films that are glued together and shaped by a laser cutter.

Inkjet printing techniques are based on different kinds of fine powders such as plaster or starch. After a layer of the powder has been dispensed by a piston, the parts of this layer belonging to the 3D object are bonded by an adhesive liquid deposited by another piston. Inkjet printing techniques can also be used to generate a 3D scaffold with different types of tissue by printing living cells and biomaterials simultaneously (10,11).

Some fabrication techniques use two materials for the construction of parts. The first material is the part material and the second is the support material (to support overhanging features during construction); the support material is later removed by heating or dissolved with a solvent or water. This is not required in techniques where a powder bed provides the support such as in SLS and inkjet printing techniques. Depending on the fabrication technique, it is also possible to combine materials of different elasticities or colors in one model. This can be useful in creating more realistic models for educational or research purposes, or for naturally looking prostheses.

5.1.3 *Microvalve-Based Bioprinting*

Microvalve-based bioprinting is a drop-on-demand technology (12). A typical microvalve-based bioprinting system consists of a three-axis movable robotic platform and an array of multiple electromechanical microvalve printheads. Each microvalve printhead is connected to an individual gas regulator that provides the pneumatic pressure (positive pressure) and the valve opening time (minimum 0.1 *ms*) which is controlled by the movement of both the plunger and the solenoid coil. The applied voltage pulse induces a magnetic field in the solenoid coil that opens the nozzle orifice by pulling the plunger up in an ascending motion. The bioink is deposited when the pneumatic pressure overcomes the fluid viscosity and surface tension at the opened orifice. The material deposition process is highly dependent on the nozzle diameter, the viscosity and surface tension of the bioink, the pneumatic pressure and the valve opening time.

5.2 3D Printed Organ Models for Surgical Applications

Medical errors are a major concern in clinical practice, suggesting the need for advanced surgical aids for preoperative planning and rehearsal (13). Conventionally, computer tomography (CT) and magnetic resonance imaging (MRI) scans, as well as 3D visualization techniques, have been utilized as the primary tools for surgical planning. These techniques are effective, but it would be useful if additional aids could be developed and utilized particularly in complex procedures involving unusual anatomical abnormalities that could benefit from tangible objects providing spatial sense, anatomical accuracy, and tactile feedback.

Recent advancements in 3D printing technologies have facilitated the creation of patient-specific organ models with the purpose of providing an effective solution for preoperative planning, rehearsal, and spatiotemporal mapping.

5.2.1 *Organ Bioprinting*

Organ printing is in general a computer-aided, dispenser-based, three-dimensional tissue engineering technology directed towards the construction of functional organ modules and eventually entire organs, layer-by-layer.

A plurality of multicellular bodies can be arranged in a pattern and allowed to fuse to form an engineered tissue (14). Such three-dimensional constructs can be assembled by printing the multicellular bodies and filler bodies.

As a shaping device, a capillary pipette can be used that is part of a printing head of a bioprinter.

The inclusion of extracellular matrix components, such as gelatin or fibrinogen, in the cell paste may facilitate the production of a multicellular body in a single maturation step, because such components can promote the overall cohesivity of the multicellular body.

Several examples for the preparation of multicellular bodies and of tissue engineering have been presented in detail (14).

Recently, human organ 3D bioprinting has attracted a lot of attention. Organ 3D bioprinting is a sophisticated procedure taking profound scientific and technological knowledge.

One of the goals is to build implantable branched vascular networks in a predefined 3D construct. Some of the recent achievements of 3D bioprinting for three large organs, such as the bone, liver and heart, have been reviewed (15).

5.2.1.1 *Bioprinting Techniques for Organ Manufacturing*

Different bioprinting techniques for organ manufacturing have been compared, as shown in Table 5.2.

Table 5.2 Bioprinting techniques for organ manufacturing (15).

Technique	Advantage	Disadvantage
Inkjet based	High printing resolution (~20 μm); several thermosensitive hydrogels can be printed; simple sample-loading requirements; low viscosity of cell suspensions (up to 106 cells/ml) or cell-laden hydrogels (3 $mP\,s$ – 30 $mP\,s$); middle cell viability (> 70%).	Limited materials can be used; complex 3D constructs are difficult to achieve; limited height (< 10 μm); potential cell desiccation; high shear stress endured by cells; droplet instability at high printing speed; poor cell sedimentation effects; poor mechanical properties.
Extrusion based	Easily updated soft and hardware; flexible geometric shapes; multiple biomaterials including cell types can be incorporated; homogeneous and heterogeneous structures can be created; good cell sedimentation effect; high cell viability (> 98%).	Material viscosity and temperature-dependent; high viscosity hydrogels may affect cell activities.

Table 5.2 (cont.) Bioprinting techniques for organ manufacturing (15).

Technique	Advantage	Disadvantage
Laser-assisted	Relatively high printing resolution (~40 μm); wide range of printable viscosity; high cell viability (>90%).	High cost; low efficiency; difficult to incorporate multiple bioactive agents; poor cell sedimentation effects; poor cell homogeneity.
Stereolithography based	Cytotoxicity of the laser beam and photoinitiators; additional post-curing process may be necessary to remove the unpolymerized liquid resin; poor cell sedimentation effects.	
Microvalve based	Relatively high printing resolution (~150); low viscosity of hydrogels (1 mPs to 70 mPs); middle cell viability (> 80%); middle cell sedimentation effect.	High shear stress suffered by cells; weak mechanical properties.

5.2.2 *Materials*

5.2.2.1 *Rigid plastics*

The early 3D printed organ models were mainly fabricated using commercial rigid plastics, primarily acrylonitrile-butadiene-styrene and poly(lactic acid) (PLA) thermoplastic filaments for FDM printing, or rigid photopolymers and resins for PolyJet technology, such as the Vero™ family of photopolymers from Stratasys.

Such models are still popular due to their accuracy in representing the anatomy of the patient with a relatively low cost (16).

Rigid plastics are polymers with high crosslinking densities and/or high molecular weights with a glass transition temperature above the room temperature. These materials are mechanically rigid with a high-impact strength and hardness. The Young's moduli for most of the rigid plastic materials are close to or within the range of *GPa*. This is at least three orders of magnitude higher than the modulus of a soft organ tissue. This discrepancy in the elastic properties of the materials relative to the organs themselves limits the direct application and realism of these models for surgical rehearsal (17,18).

Also, these organ models have demonstrated an utility in a variety of medical fields, including cardiology, urology, neurology, and hepatology (13).

For urology purposes, a 3D printed patient-specific, cancerous kidney model with accurate anatomy for applications in urological oncology was fabricated (19). Here, as a transparent flexible material, Materialise HeartPrint™ Flex was used as the main cortex and the Vero family of rigid photopolymers in different colors as the remaining structures. Such kidney models with tumor sections allowed the surgeons to evaluate the complexity of the tumor and its positional relationship with respect to other parts of the organ. This model facilitated the operational planning for partial nephrectomy or ablative therapy. In a real surgical case, the model was used to assist surgeons in the selection of an approach for partial nephrectomy, as well as for resection guidance during the surgery.

Also, a 3D printed kidney graft and pelvic cavity model was constructed using the PolyJet printing process (20). Here, mainly the Tango family of photopolymers with different colors was used. The model was successfully applied for preoperational planning and accurate simulation of the surgical procedure for kidney transplantation.

These materials are available in a variety of colors, including clear, magenta, and cyan. They can be used to distinguish different printed sections and anatomical details of organ models. The above-mentioned 3D printed models were found to be helpful for recognizing the anatomical features of the organs in the corresponding operations (18).

For neurological applications, 3D printed hollow intracranial aneurysm models with rigid walls using an FDM process and PLA filaments were tested (21). The models could accurately replicate the aneurysm anatomy of the patients in a digital subtraction angiography image. For this reason they could be used for surgical aid applications as well as for MRI flow phantoms and computational fluid dynamic studies based on rigid models for simulation of aneurysm hemodynamics.

5.2.2.2 *Elastomers*

Beyond conventional rigid plastics, there is a possibility of 3D printing of elastomeric, i.e., rubber-like, and flexible materials.

Examples of such materials are the Tango™ family (Stratasys) of photopolymers for PolyJet printing, or thermoplastic elastomer (TPE) filaments such as NinjaFlex® (NinjaTek), SemiFlex™ (NinjaTek), and PolyFlex™ (Polymaker) for FDM printing.

In contrast to TPE filaments with elastic properties, rubbers are thermoset polymers with network structures. These thermoset network polymers are not suitable for FDM printing because the polymer chain motion is greatly restricted by a high degree of crosslinking after heating, such that they cannot be remanufactured after their initial heat forming.

5.2.2.3 *Powder-Based Materials*

Powder-based materials such as starch, cellulose, and plaster powder can also be solidified with binding materials in the course of inkjet 3D printing. These materials have also been evaluated for the fabrication of different organ models for surgical applications.

Despite the mismatch of their mechanical properties with real organs, such models can provide accurate anatomical details with convenient low-cost fabrication.

In the field of cardiology, an example was demonstrated for creating 3D printed cardiac models using a starch/cellulose powder, a polymer as binder, and an elastomeric urethane resin for further infiltration (22).

There, the participating surgeons could detect the bypass grafts and their position with respect to the sternum from the model. The sterilized model was further used in the operating room for guiding the intraoperative procedures for reopening the sternum.

In another example, a 3D printed cardiac model was shown using plaster-based powder with a binding material (23). The final model allowed surgeons to obtain intracardiac views that are difficult to achieve during the actual operation. Additionally, the plaster materials were used in neurology applications.

Using plaster materials, 3D printed models were developed of a skull base and intracranial tumors (24,25). Such models can provide a better visualization of the anatomy and size of the organ, tumor, and their positional relationships. The models were further used to provide realistic surgical practice and sensation via insertion of surgical instruments under microscopic observation.

5.2.2.4 *Bone Repair Materials*

In recent years, a variety of 3D printing technologies have been further developed to construct bone repair materials that mimic the composition of bone tissues and the microenvironment of bony extracellular matrices (15).

A hard tissue repair material using titanium and caprolactone has been described (26). Also, a porous calcium phosphate cement/alginate scaffold has been described by depositing a solution of α-tricalcium phosphate-based powder and sodium alginate in a calcium chloride bath (27).

A single nozzle low-temperature rapid prototyping (RP) technology was used to prepare large bone repair materials with predefined (go-through) channels 200–500 μm in diameter, which were hard to produce using traditional manufacturing technologies (28).

In 2016, an elastic construct for bone regeneration was developed. Poly(caprolactone) (PCL) or poly(lactic-*co*-glycolic acid) (PLGA) and hydroxyapatite was dissolved in a trisolvent mixture as bioink. The printed 3D constructs can be handled in a versatile manner, such as cutting, folding, rolling and suturing. Human mesenchymal stem cells seeded on the 3D constructs showed a significant up-regulation of pro-osteogenic genes, collagen type I, osteocalcin, and osteopontin at day 28. When the 3D constructs were implanted in a macaque calvarial defect for 4 weeks, excellent new bone formation accompanied by the vascularization and integration of surrounding tissue was observed (29).

5.2.2.5 *Stimuli-Responsive Polymers*

Synthetic polymers possess more reproducible physical and chemical properties than their naturally occurring counterparts (30). They have also emerged as an important alternative for fabricating tissue substitutes because they can be molecularly tailored to have a vast array of molecular weights, block structures, active functional groups, and mechanical properties.

A synthetic polymer has been developed using controlled living radical polymerization. It can be printed into well-defined structures. The polymer has a low cytotoxicity before and after printing.

The incorporation of gelatin-methacrylate coated PLGA microparticles within the hydrogel provided cell adhesion surfaces for cell proliferation. The results indicated a possible application of the microparticle seeded, synthetic hydrogel as a direct printable tissue or organ substitute (30).

Stimuli-responsive hydrogels exhibiting physical or chemical changes in response to environmental conditions have attracted growing attention (31).

Poly(*N*-isopropyl acrylamide) is a temperature-responsive hydrogel. It has been extensively studied in various fields of science and engineering. However, manufacturing of poly(*N*-isopropyl acrylamide) has been heavily relying on conventional methods such as molding and lithography techniques that are inherently limited to a 2D space.

The 3D printing of poly(*N*-isopropyl acrylamide) has been reported using a high-resolution digital additive manufacturing technique, projection micro-stereolithography (31).

Projection micro-stereolithography is a lithography-based additive manufacturing technique that is fast, inexpensive, and flexible in material selection (32,33). Here, a 3D model is generated using computer-aided design software and digitally sliced into a series of cross-sectional images of the 3D model. Each digital image is transferred to a digital mask to optically pattern UV light, which is then projected through a reduction lens and focused on the surface of photocurable resin (31). The patterned UV light converts liquid resin to a solid layer through photopolymerization. Once a layer is formed, the linear stage drops the sample holder on which the object is built in order to introduce fresh liquid resin for the next layer. The subsequent layer is polymerized in the same manner on top of the preceding layer. This process is repeated until all the layers are complete to build the 3D object.

A control of the temperature-dependent deformation of 3D printed poly(*N*-isopropyl acrylamide) can be achieved by controlling the parameters of the manufacturing process, as well as polymer resin composition.

A sequential deformation of the 3D printed poly(*N*-isopropyl acrylamide) structure could be demonstrated by the selective incorporation of an ionic monomer (methacrylamidopropyltrimethyl ammonium chloride) that shifts the swelling transition temperature of the poly(*N*-isopropyl acrylamide).

When the positively charged ionic monomer is integrated into the crosslinked network the hydrophilicity of a polymer chain is increased, because of the growing ratio of cationic sites (34,35).

This fast, high-resolution, and scalable 3D printing method for stimuli-responsive hydrogels may enable a lot of new applications in diverse areas, including flexible sensors and actuators, biomedical devices, and tissue engineering (31).

5.2.2.6 *Tissue-Mimicking Materials*

Tissue-mimicking materials can be both biopolymers, e.g., gelatin, gellan gum, agar, and agarose, and synthetic polymers, e.g., poly(urethane) (PU), poly(vinyl alcohol) (PVA), poly(vinyl chloride)

(PVC), silicones that are vulcanized at room temperature, and poly-(dimethyl siloxane) (36).

The properties of these tissue-mimicking materials and tissues have been detailed, including elastic modulus, needle insertion friction force, speed of sound, acoustic attenuation, and visible light transmittance (36).

These materials have been used in various arenas of medicine for simulation purposes, including medical imaging modalities (37–39), cardiac strain estimation (40, 41), thermal therapy (42, 43), and surgical simulation and training (44, 45).

The composition of these materials can be tailored to replicate the specific properties of soft tissue depending on the application. For example, phantom organs based on polymers, such as gelatin, agar, PVC, and PVA, have been developed to mimic the acoustic properties of soft tissue, including the speed of sound, acoustic impedance, attenuation, and backscattering coefficient. These devices were utilized in ultrasound imaging for system calibration, development of new techniques, and the training of technicians (13).

For developing organ models with implications in surgical planning and training, the composition of the selected material should be modified to closely match the mechanical properties, including elastic modulus, viscoelastic behavior, hardness, and ultimate strength, of the biological soft tissue.

Models fabricated with such materials exhibit a more accurate haptic feedback and mechanical behavior, which is analogous to the real organ. A common technique for incorporating tissue-mimicking materials in organ models is to first use 3D printing to create a mold. Then a casting occurs with tissue-mimicking materials.

These molds can be created using one of the following approaches (13):

1. 3D printing a negative mold of the organ and infusing it with the tissue-mimicking material, or
2. Employing an approach similar to lost-wax casting, i.e., directly 3D printing the organ model using commercially available materials.

This so produced 3D printed model is then used as a template for creating a mold (for example, via silicone molding methods) and the mold cavity is subsequently filled with the tissue-mimicking material to fabricate the final organ model (46–50).

Although molding techniques provide a platform for using customized tissue-mimicking materials and fabricating organ models, they fall short in different aspects that hamper their widespread adaptation for clinical practice. These mold-based fabrication procedures typically involve several steps, which could be time-, labor-, and cost-intensive. However, they are also prone to the introduction of inaccuracies in the final model (13).

Poly(vinyl chloride). A factorial experimental design was created to study the effects of the ratio of softener to PVC polymer, mass fraction of mineral oil, and mass fraction of glass beads on mechanical and medical imaging properties of soft PVC. Factorial analysis was used to identify the statistical significance of these factors and a regression model was developed to find the quantitative relationship between PVC properties and the three factors. A regression model was used to find values of three factors of the PVC as an example to achieve targeted mechanical and medical imaging properties and validate the results experimentally (36).

Based on this approach, PVC phantom tissues for clinical simulators or medical research devices could be designed with desired mechanical properties and an appropriate composition of PVC could be identified using the regression model. The medical imaging properties of the PVC samples have narrow ranges and commonly are not close enough to those of real tissues. The regression models of medical imaging properties were able to predict the value of these properties of the PVC sample with a known composition (36).

5.2.3 *Liver*

The liver is a vital visceral organ in the human body (15). Unlike a structural organ bone, liver 3D bioprinting has several bottleneck problems to solve: One of them is how to construct the branched vascular and bile duct networks, another of them is how to distribute more than three cell types in a predefined 3D construct with a high cell density and make them develop into functional tissues (51).

Cell-laden gelatin-based hydrogels were assembled into large scaled-up liver tissues with predefined structures (go-through channels) using a extrusion-based 3D printing system under the instruction of an experiential CAD model (52–55).

The predefined structures were printed via a pressure-controlled syringe. This technique allows the deposition of cell-laden hydrogel solutions with high concentration and velocity. Cylindrical channels with diameters ranging from 100 to 300 μm were produced. After 3D printing, the gelatin-based polymers in the cell-laden constructs were submitted to a chemical crosslinking process to stabilize the structures and improve the mechanical properties. Hepatocytes encapsulated in the gelatin-based hydrogels remained viable and produced hepatic ECMs during an *in-vitro* culture for 8 weeks.

A large scaled-up vascularized liver tissue was first produced using another experiential CAD model (56, 57). Afterwards, actual bioartifical organ manufacturing was put forward and developed very quickly.

A 3D printed complicated organ with a completely confluent endothelial layer that covered the inner channels of the vascular network was produced. It was possible to observe that endothelial cells aligned inside the surface of the predefined channels. More than three cell types formed functional tissues in a complex 3D construct. This technique has advanced in organ manufacturing areas (58, 59).

5.2.4 *Heart*

The heart is one of the most important internal organs of human beings. It is composed of three different cardiac tissues:

1. Myocardium,
2. Pericardium, and
3. Endocardium.

The myocardium is the thick muscular layer of the heart wall which consists of cardiomyocytes, aligning themselves in an anisotropic manner and promoting the electrical activation of the cardiac muscles, and taking up to 30%–40% of the entire cell population.

The pericardium is a conical, flask-like, double-wall fibroserous sac that encloses the blood vessels from the root of the heart.

The endocardium is the endothelial lining of the innermost heart chambers and heart valves. It is primarily made up of endothelial cells that seal the heart and connect the surrounding blood vessels (60,61).

The rest of the cell types of the heart are mainly non-myocyte fibroblasts (62). The elasticity of the cardiomyocytes and their collagen-based extracellular matrices in a normal heart are pliable and tough enough to generate actomyosin forces and pump the heart.

A stream of cell-laden hydrogel microparticles in a well-defined topological pattern formed 3D myocardial patches using a inkjet-based bioprinting technique (63). This technique is supported by the self-assembly and self-organizing capabilities of cells.

Patterned human stem cells and endothelial cells with laser printing for cardiac regeneration have been reported (64). Also, human cardiomyocyte progenitor cells in a alginate hydrogel were printed (65). The human cardiomyocyte progenitor cells in the alginate hydrogel showed an increase of cardiac commitment while at the same time maintaining viability and proliferation.

Trileaflet valve-like conduits were constructed using sinus smooth muscle cells and alginate/gelatin hydrogels (66). The cell viability in the alginate/gelatin hydrogels attained 81.4%.

Then, a similar study was carried out in the same group using human aortic vascular interstitial cells in methacrylated hyaluronic acid or gelatin methacrylate hydrogels (67). High human aortic vascular interstitial cell viability of the encapsulated cells are greater than 90%. Promising remodeling potentials were obtained using this technology. The main concern of this technology is that poly-(methacrylate) is a non-biodegradable polymer. It may hinder the cells formation of functional tissues during the later cultures.

In 2015, a heart CAD model was created using freeform reversible embedding of hydrogel (68). An extrusion-based 3D bioprinting technology was used to produce a functional cardiac tissue, and particularly a semilunar heart valve with three main components: a relatively stiff heart valve root populated by contractile smooth muscle cells, three thin flexible leaflets containing fibroblastic interstitial cells and three sinuses (69). The semilunar heart valve allows blood to be forced into the arteries and prevents the backflows. Hybrid hydrogel properties were studied by changing concentrations of the two compositions: methacrylated hyaluronic acid and gelatin methacrylate. The optimized hydrogel formulation was mixed with human aortic vascular interstitial cells. After 7 *d* in static culture, the

3D bioprinted valve showed well-maintained structure, high viability of the encapsulated cells (90%), as well as promising remodeling potentials.

5.2.5 *Cartilage*

Cartilage is a resilient and smooth elastic organ of the body, which protects the ends of long bones at the joints, e.g., the elbows, knees and ankles. It is a structural component made up of specialized cells called chondrocytes. They are in the rib cage, the ear, the nose, the bronchial tubes or airways, the intervertebral discs, and many other body components. The chondrocytes produce large amounts of extracellular matrices composed of proteoglycan, collagen and elastin fibers.

Specifically, there are no blood vessels in cartilage to supply the chondrocytes with nutrients. It is not as hard and rigid as bone, but it is much stiffer and much less flexible than muscle. Like many other organs, cartilage exhibits multiple zonal organizations with highly coordinated cell distribution. Cartilage can be categorized into three types (15):

1. Hyaline cartilage with low-frication and wear-resistant properties,
2. Elastic cartilage with flexible property, and
3. Fibrocartilage with tough and inflexible properties.

Due to the lack of blood vessels, cartilage grows and repairs more slowly than other tissues/organs.

3D printing technologies, which have been used in cartilage regeneration, emerged relatively late (15).

A printed meniscus scaffold has been reported with two different zones: the white zone, which is located at the inner zone of the meniscus, consists of chondrocyte-like cells with abundant collagen type II and glycosaminoglycans, whereas the red zone, which is in the other zone of the meniscus, contains fibroblast-like cells with collagen type I (70). Human connective tissue growth factor and transforming growth factor $\beta 3$ were then placed in the red and white zones respectively. The two zones spatiotemporally released the growth factors and induced the human synovium smooth muscle cells to form a zone-specific matrix, i.e., collagen type II in the white zone and collagen type I in the red zone. The zone-specific phenotypes were further exhibited in a 3-month implantation of a sheep partial meniscectomy model.

In 2015, a printed hybrid cartilage construct was reported containing chondrocyte, alginate, and PCL (71). Then, an autologous cartilage construct was developed consisting of autologous chondrocyte, alginate, and PCL for auricular reconstruction (72). The synthetic PCL was printed with alginate hydrogel and cells, which can provide the construct with long-term stability. The rigid properties of PCL may induce abrasion of the surrounding cartilage tissue.

A biodegradable PU was fabricated involving a cartilage construct which exhibited a high strain recovery property (73). Other bioactive compounds, such as hyaluronic acid and growth factors, can be encapsulated into the PU bioink and induce high glycosaminoglycans secretion at 4 weeks after implantation into rabbit osteochondral defects. The formation of cartilage was observed by safranin-O staining.

Conventional approaches for the regeneration of the damaged tissue based on integrated manufacturing are limited by their inability to produce precise and customized biomimetic tissues (74). On the other hand, 3D bioprinting is a promising technique with increased versatility because it can co-deliver cells and biomaterials with proper compositions and spatial distributions.

The recent progress has been reviewed concerning the complete 3D printing process involved in functional cartilage regeneration, including printing techniques, biomaterials and cells (74).

Also, the combination of 3D *in-vivo* hybrid bioprinting with spheroids, gene delivery strategies and zonal cartilage design have been discussed. This may be a future direction in the field of cartilage regeneration (74).

5.2.6 *Bionic Ears*

Because it is possible to interweave biological tissue with functional electronics the creation of bionic organs can be achieved, which possess enhanced functionalities over their human counterparts (75).

A novel strategy was presented that uses the additive manufacturing of biological cells with structural- and nanoparticle-derived electronic elements.

The concept has been exemplified by the construction of a bionic ear by 3D printing of a cell-seeded hydrogel matrix in the anatomic geometry of a human ear, along with an intertwined conducting polymer consisting of infused silver nanoparticles.

This printed ear exhibits enhanced auditory sensing for radio frequency reception, and complementary left and right ears can listen to stereo audio music (75).

5.2.7 Skin

The skin is the largest organ of the human body, which accounts for about 15% of the weight of the body and maintains the body's temperature through sweat or other mechanisms (76).

Along with sweat glands, the skin contains oil glands to keep the skin from drying out and the hair from becoming brittle. The skin consists of three layers, namely the epidermis, dermis and hypodermis. The epidermis is the outer layer, consisting of keratinocytes (KCs), the dermis is the middle layer, consisting of collagen and fibroblasts, and the hypodermis is the inner layer, consisting of lipocytes and collagen. There are about 19 million skin cells in every square inch of the human body! Although numerous studies have tried to generate full-thickness skin substitutes, most methods are dependent on the technique that seed cells on a porous scaffold, with which it is not easy to recapitulate the heterogeneity of skin comprising multiple types of cells. 3D bioprinting allows similar skin geometry to be built via the spatiotemporal pattern of the related cell types of the skin (77).

Traditional skin substitutes either are made of natural or synthetic polymers, which could promote skin tissue regeneration to a certain degree. These substitutes have been used in surgical therapies when an autologous flap is not desirable. However, these substitutes have not been successfully used in clinical applications due to some technological limitations such as the lack of multilayer structures, vascularization and innervation (78).

In 2006, Ringeisen *et al.* printed living cells for skin regeneration using a laser-assisted technique (79). The process uses radiation pressure from the scattering of energetic photons in a laser beam to deposit cell solutions in a high concentration and a rapid velocity of 10 $m\,s^{-1}$, and micrometer resolution. Multiple skin cells were deposited with micron-scale resolution from a transfer layer or reservoir.

Human fibroblasts were delivered using a piezoelectric drop-on-demand inkjet printing technique (80).

An extrusion-based printing system was used to fabricate skin substitutes using collagen, fibroblasts and keratinocytes (81). A printed layer of cell-containing collagen was crosslinked by coating the layer with nebulized aqueous sodium bicarbonate. The process was repeated in layer-by-layer fashion on a planar tissue culture dish, resulting in two distinct cell layers of inner fibroblasts and outer keratinocytes.

Recently, a 3D printed full-thickness skin substitute containing dermis and epidermis layers was reported (82). A mixture of gelatin and fibrinogen was used as bioink. After 26 *d* of culture, the 3D printed skin substitute exhibited similar histological characteristics as human skin. Not only the main skin tissues but also the skin appendages, such as sweat glands, has been mimicked (83).

A 3D matrix was bioprinted as the restrictive niche for direct sweat gland differentiation of epidermal progenitors by different pore structure of 300 μm or 400 μm nozzle diameters printed (84).

It was found that a long-term gradual transition of differentiated cells into glandular morphogenesis occurs within the 3D construct *in-vitro* (84). At the initial 14 *d* culture, an accelerated cell differentiation was achieved with inductive cues released along with gelatin reduction. After the protein release was completed, the 3D construct guides the self-organized formation of sweat gland tissues, which is similar to that of the natural developmental process.

However, a glandular morphogenesis was only observed in 300 μm printed constructs. In the absence of the 3D architectural support, the glandular morphogenesis did not occur. This striking finding allowed the identification of a previously unknown role of the 3D printed structure in glandular tissue regeneration, and this self-organizing strategy can be applied to forming other tissues *in-vitro* (84).

5.2.8 Scaffolds

Traditional techniques of scaffold fabrication include (85):

- Fiber bonding,
- Solvent casting and particulate leaching,
- Membrane lamination,
- Melt molding, and
- Gas foaming.

These techniques have several drawbacks such as the extensive use of highly toxic organic solvents and long fabrication periods. Therefore, more recent rapid prototyping techniques have been utilized by tissue engineers to produce three-dimensional porous scaffolds (85). Rapid prototyping technologies for scaffold fabrication are summarized in Table 5.3.

Table 5.3 Rapid prototyping technologies for scaffold fabrication (2).

Technology	References
3D plotting and direct ink writing	(86)
Laser-assisted bioprinting	(87)
Selective laser sintering	(88)
Stereolithography	(89)
Fused deposition modeling	(90, 91)
Robotic-assisted deposition/robocasting	(92)

3D printing technologies allow the design and fabrication of complex scaffold geometries with a fully interconnected pore network. A blend of starch-based polymer powders, such as cornstarch, dextran and gelatin, was developed for a 3D printing process. Cylindrical scaffolds with five different designs were fabricated and post-processed to enhance the mechanical and chemical properties (85).

5.2.8.1 Hydrogels

An attractive area of research is 3D printing of hydrogels, because it is capable of fabricating intricate, complex and highly customizable scaffold structures that can support cell adhesion and promote cell infiltration for tissue engineering (93). However, pure hydrogels alone lack the necessary mechanical stability and are too easily degraded to be used as printing ink. To overcome this problem, significant progress has been made in the 3D printing of hydrogel composites with improved mechanical performance and biofunctionality.

Existing hydrogel composite 3D printing techniques, including laser-based 3D printing, nozzle-based 3D printing, and inkjet printer-based 3D printing systems, have been reviewed (93).

Four main hydrogel composite systems are detailed in this review: polymer- or hydrogel-hydrogel composites, particle-reinforced hydrogel composites, fiber reinforced hydrogel composites, and anisotropic filler-reinforced hydrogel composites.

Hydrogels for 3D printing can be divided into protein-based natural hydrogels such as gelatin, collagen, and silk, or polysaccharide-based natural hydrogels such as chitosan, agarose, hyaluronic acid, alginate, and cellulose, or synthetic hydrogels such as poly(ethylene glycol) (PEG), poly(urethane), and poly(acrylamide).

Alginate, a common hydrogel crosslinked by ionic interactions or phase transition, has been widely used in the field of soft tissue engineering due to its biodegradability and low toxicity. However, alginate limits the cellular adhesion due to the lack of adhesion sites for the cells.

The incorporation of micro- or nanoparticles into a hydrogel is widely used to enhance the mechanical and biological properties of pure hydrogels due to their low cost, ease of preparation, and isotropic strengthening behavior.

Biphasic calcium phosphate microparticles in the range of 106 μm – 212 μm were used for composite reinforcement and Matrigel® or alginate as the hydrogel matrix (94). This particle-reinforced hydrogel composite was implanted in bone defects and used as an osteoinductive bone filler.

A fiber reinforcement can also improve the mechanical properties of a hydrogel matrix in which the fiber contents and its distribution inside its matrix determine mechanical properties such as stiffness and strength of the composites (95–97). In the case of conventional 3D printing systems, short fiber reinforcements are the most commonly used due to their easy processing procedure at low cost. The fibers can be directly incorporated into the hydrogel matrix via simple mixing and transferred into the syringe for printing.

Anisotropic filler-reinforced hydrogel composites can be used for 3D printing. Nanoclay is a nanoparticle which is composed of layered hydrous silicate. It has been used in a wide range of applications, such as pharmaceuticals, paints, and cosmetics as well as catalysis, owing to their good surface properties and excellent rheology controllability. Depending on the type of clay, each layer consists of two or more sheets of either $(AlO_3(OH)_3)_6$ octahedra or $(SiO_4)_4$ tetrahedra (93).

5.2.9 *Personalized Implants*

The 3D printing technologies not only enable the personalization of implantable devices with respect to patient-specific anatomy, pathology and biomechanical properties but they also provide new opportunities in related areas such as surgical education, minimally invasive diagnosis, medical research and disease models (98).

The recent clinical applications of 3D printing have been reviewed, with a particular focus on implantable devices. The current technical bottlenecks in 3D printing in view of the needs in clinical applications are explained and recent advances to overcome these challenges have been presented.

Bioprinting is an exciting subfield of 3D printing. This method is covered in the context of tissue engineering and regenerative medicine using bioinks. Also, emerging applications of bioprinting beyond health, such as biorobotics and soft robotics, were discussed (98).

5.2.10 *Neural Tissue Models*

The recent developments in 3D *in-vitro* neural tissue models have been reviewed (99). There has been a particular focus on the emerging bioprinted tissue structures. Specific examples to describe the merits and limitations of each model have been given in terms of different applications.

Bioprinting offers a revolutionary approach for constructing repeatable and controllable 3D *in-vitro* neural tissues with diverse cell types, complex microscale features and tissue level responses. Further advances in bioprinting research would likely consolidate existing models and generate complex neural tissue structures bearing higher fidelity, which is ultimately useful for probing disease-specific mechanisms, facilitating development of novel therapeutics and promoting neural regeneration (99).

5.2.10.1 *Polymer Molecule Alignment*

Proper cell-material interactions are critical to cell function and thus successful tissue regeneration (100). A lot of fabrication processes have been developed to create microenvironments to control cell attachment and organization on a 3D scaffold. However, these approaches often involve heavy engineering and only the surface layer can be patterned.

Poly(lactic-*co*-glycolic acid) (PLGA) with lactic acid/glycolic acid in a ratio of 85:15 and a molecular weight of 35 *kD* was used. A 3D printed scaffold was fabricated using the 3D-Bioplotter (EnvisionTEC, Gladbeck, Germany) with a direct melt extrusion technique. The scaffold was programmed with inner patterns using the provided EnvisionTEC software. For printing, the material was loaded into the printing cartridge and melted at 165°C and extruded at 9 *bar* with an average speed of 1.5 $mm\,s^{-1}$ using a 0.2 *mm* inner diameter needle based on previously established methods (101).

A parallel pattern scaffold has a fiber diameter of 0.2 *mm* parallel to each other with 0.2 *mm* edge-to-edge spacing of two adjacent fibers. For random pattern, the angle to the contour and the spacing of each layer were randomly selected using a random generator package in R software. All scaffolds have a dimension of 4 *mm* (length) × 4 *mm* (width) × 1.5 *mm* (height). The casted PLGA was made by melting the raw PLGA material and shape to the same size as the printed scaffold (100).

It was found that 3D extrusion-based printing at high temperature and pressure will result in an aligned effect on the polymer molecules. This molecular arrangement will further induce the cell alignment and different differentiation capacities. In particular, articular cartilage tissue is known to have zonal collagen fiber and cell orientation to support different functions, where collagen fibers and chondrocytes align parallel, randomly, and perpendicular, respectively, to the surface of the joint. Therefore, the cell alignment was evaluated in a cartilage model.

Small angle X-ray scattering analysis was used to substantiate the polymer molecule alignment phenomenon. The cellular response was evaluated both *in vitro* and *in vivo*. Seeded mesenchymal stem cells showed a different morphology and orientation on scaffolds, as a combined result of polymer molecule alignment and printed scaffold patterns. Gene expression results showed an improved superficial zonal chondrogenic marker expression in a parallel-aligned group. The cell alignment was successfully maintained in an animal model after 7 *d* with distinct mesenchymal stem cell morphology between the casted and parallel printed scaffolds.

This 3D printing induced polymer and cell alignment will have a significant impact on developing a scaffold with controlled cell-material interactions for complex tissue engineering while avoiding complicated surface treatment. Therefore it provides a new concept for effective tissue repairing in clinical applications (100).

An original 3D bioprinting method has been developed by modifying an exceptional 3D printer (102). Using a composite material, bioprinting was carried out to create the ideal scaffold material to contribute to the regeneration of a certain amount of tissue types in humans.

After bypassing the extruder and the heating system of the 3D printer, instead of using solid filaments, a polymer-ceramic composite was dissolved using an organic agent and the bioprinting was done. During the bioprinting, the dissolving agent evaporated quickly and so the solidification process was completed.

Despite traditional 3D printing benefitting from the glass transition temperature of the materials, regardless of the temperature, the rapid prototyping technology has been merged with controlled flow rate of the composite solution and evaporation of the solvents were adjusted meticulously for proper solidification and layer-by-layer bioprinting of the scaffolds (102).

5.2.10.2 *Scaffolds for Biomineral Formation*

The synthesis of nylon-12 scaffolds by 3D printing has been presented (103). Also, their versatility as matrices for cell growth, differentiation, and biomineral formation has been demonstrated.

The porous nature of the printed parts makes them ideal for the direct incorporation of preformed nanomaterials or material precursors, leading to nanocomposites with very different properties and environments for cell growth.

Additives such as tetraethyl orthosilicate derivatives applied at a low temperature promote successful cell growth, partly due to the high surface area of the porous matrix.

The incorporation of presynthesized iron oxide nanoparticles led to a material that showed rapid heating in response to an applied magnetic field, an excellent property for use in gene expression and, with further improvement, a chemical-free sterilization.

These methods also avoid changing polymer feedstocks and contaminating or even damaging commonly used selective laser sintering printers. The chemically treated 3D printed matrices have great potential for use in addressing the current issues surrounding bone grafting, implants, and skeletal repair, and a wide variety of possible incorporated material combinations could impact many other areas (103).

5.3 Bioinks

The recent issues concerning bioinks have been reviewed (104, 105). A bioink is a particular suspension of cells or tissue spheroids in a liquid solution (106).

The biology of a cell is influenced by chemical and physical cues of the surrounding environment (107, 108).

Cells cultured in 2D and 3D systems demonstrate a significantly different behavior with respect to migration, adhesion, gene expression and mitosis (109, 110).

3D culture constructs are important for mimicking native cell tissue *in vitro* (111) and recently bioinks have emerged as a candidate for reliable and fast 3D bioprinted cell culture systems. Bioinks are composed of cells suspended in a liquid, pre-gel solution that is then printed onto a surface or into a 3D scaffold using mechanical extrusion (106). During the printing process, the bioink solution is gelled by polymer crosslinkers, photo activation, or thermal activation while leaving the cells intact and viable. The final bioink construct is a hydrogel that physically constrains the suspended cells. Bioink hydrogels maintain cell viability, but can be tailored for specific material properties or scaffolding dimensions.

Bioinks should fulfill the following requirements for use in bioprinting (112):

- Printability: Current 3D bioprinting systems generally have a limitation over the choice of materials because of printability, as the use of nozzles and/or energy to expel the bioink

with cells limits the viscosity and surface tension (87). To keep viscosity low during the bioprinting process and to ensure structural integrity afterwards, various crosslinking methods, such as polymer crosslinking, photocrosslinking, and thermal crosslinking, are used (106,113). Furthermore, the shear-thinning ability of bioinks is ideal for enhancing cell viability after the bioprinting procedure.
- Biocompatibility: Bioink should not cause inflammatory or immune response. It should biodegrade if needed, and support cell attachments and proliferation *in situ* in a controlled or predefined manner. Some constructs are designed to biodegrade in the bioprinted or implanted area, so the degradation by-products and the construct itself should be harmless to the subject (114).
- Mechanical properties: For regenerative tissue engineering using 3D bioprinting, bioinks should provide the required strength and elasticity for mimicking the mechanical properties of native tissues and maintaining the original printed structure to support cell growth. Tensile strength and stiffness are especially important in the tissue engineering of hard tissues such as bone (115).
- Ease of spatial arrangement: Often, viscosity and other material properties of bioink determine the resolution and microscale patterning ability (116). The 3D architectural structure (dispensing single or multi-bioinks in a predefined space) is crucial for developing tissue constructs that mimic the native tissue or organ.

Various bioinks are used in 3D cell bioprinting for fabricating cell-laden tissue constructs to provide strength and protection, keep cells moist, and allow printing without nozzle clogging. The materials most predominantly used in 3D cell bioprinting are organic solutions such as alginate, hydrogels, collagen, and hyaluronic acid (111).

Synthetic polymers such as PCL are used to provide mechanical strength to the construct. Recently, nanocellulose has attracted attention due to its various characteristics. Nanocellulose materials consist of nanofibrils which mimic the fibril network of collagen. Furthermore, nanocellulose has high water content, mechanical strength and shear-thinning properties (117).

There have been some interesting recent developments with bioinks derived through tissue and organ decellularization, dehydration, and pulverization (105,118,119).

Materials that can be used in bioinks and recent studies are summarized in Table 5.4. Some of these compounds are shown in Figure 5.1.

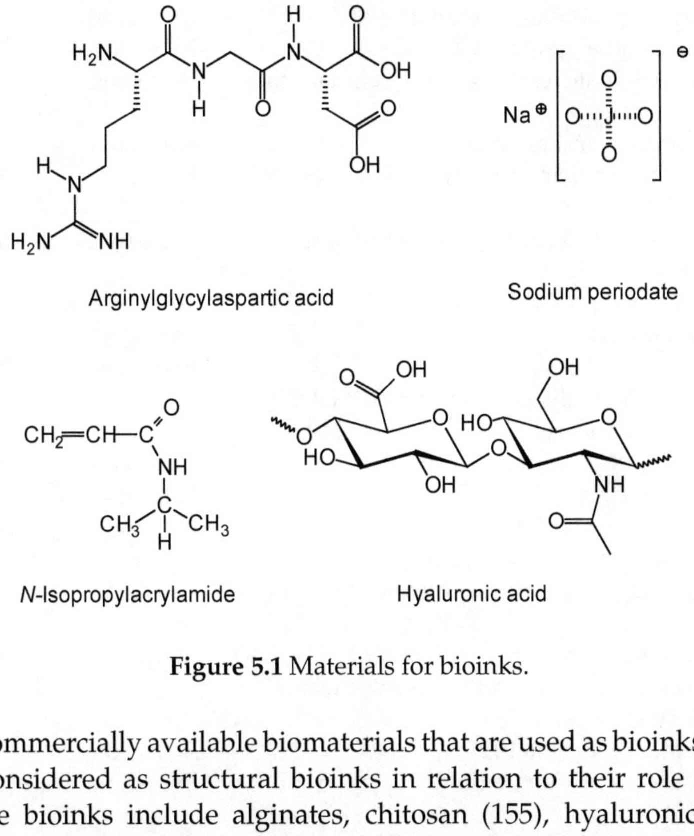

Arginylglycylaspartic acid

Sodium periodate

N-Isopropylacrylamide

Hyaluronic acid

Figure 5.1 Materials for bioinks.

Commercially available biomaterials that are used as bioinks may be considered as structural bioinks in relation to their role (105). These bioinks include alginates, chitosan (155), hyaluronic acid (156), gelatin methacryloyl (157), collagen (78), and their blends.

5.3.1 *Cytocompatible Bioink*

A cytocompatible inkjet bioprinting approach that enables the use of a variety of bioinks to produce hydrogels with a wide range of characteristics has been developed (158). The stabilization of bioinks

Table 5.4 Materials for bioinks (139).

Material	Method	Reference
Nanocellulose	Extrusion	(120)
Alginate	Extrusion	(121)
Collagen/gelatin/alginate hydrogel	Extrusion	(122)
γ-Irradiated alginate, PCL fibers	Extrusion	(123)
Gellan, alginate, cartilage extracellular matrix particles	Extrusion	(124)
M13 phages and alginate	Extrusion	(125)
Collagen, alginate, human adipose stem cells (hASCs)	Extrusion	(126)
Alginate, carboxymethyl-chitosan, and agarose	Extrusion	(127)
Commercial poly(ethylene glycol) (PEG)-based bioink	Droplet based	(128)
Gelatin-based bioinks	Extrusion	(129)
Poly(ethylene glycol) diacrylate (PEGDA), gelatin methacrylate (GelMA), eosin		(130)
Alginate, PCL/alginate mesh	Extrusion	(131)
Hyaluronic acid	Extrusion	(132)
Poly(urethane) (PU), c2c12 cells, NIH/3T3 cells, hyaluronic acid, gelatin, fibrinogen	Extrusion	(133)
PCL, collagen, and three different types of cells	Extrusion	(134)
Gelatin, poly(ethylene oxide) (PEO), HEK293 cells, human umbilical vein endothelial cells (HUVECs)	Extrusion	(135)
Acrylated, pluronic F127	Extrusion	(136)
Alginate in phosphate buffered saline (PBS), hASCs	Extrusion	(137)
Collagen/extracellular matrix (ECM) and alginate, hASCs	Extrusion	(138)
Hyaluronic acid and gelatin	Extrusion	(119)

Table 5.4 (cont.) Materials for bioinks (139).

Material	Method	Reference
Type I collagen and chitosan agarose blends, human bone marrow-derived mesenchymal stem cells (hMSCs)	Extrusion	(140)
Decellularized adipose tissue (DAT) matrix bioink, hASCs	Extrusion	(141)
Alginate, GelMA, HUVECs	Extrusion	(142)
Spider silk protein, human fibroblasts	Extrusion	(143)
Poly(*N*-isopropyl acrylamide), poly(*N*-isopropyl acrylamide) grafted hyaluronan (HA-pNIPAAM), methacrylated hyaluronan (HAMA)	Extrusion	(144)
Sodium alginate, sodium periodate, Arginylglycylaspartic acid (RGD) peptides	Extrusion	(145)
Fibroblasts, sodium alginate, poly(styrene) microbeads and 3T3 cells	Droplet based	(146)
Gelatin, methacrylic anhydride	Droplet based	(147)
Gelatin, methacrylamide, gellan gum	Extrusion	(148)
MG63 cells, alginate, PCL electrospun scaffold	Laser assisted	(149)
Polylactic acid, gelatin methacrylamide gellan gum, Mesenchymal stem cells (MSCs)	Extrusion	(150)
Alginate, gelatin, hydroxyapatite, hMSCs	Extrusion	(151)
Nanofibrillated cellulose (NFC), alginate	Extrusion	(152)
Various natural and synthetic materials such as PEG and gelatin	Extrusion	(153)
Silk fibroin, gelatin, Human turbinate mesenchymal stromal cells (hTMSCs)	Extrusion	(154)
Decellularized adipose (adECM), cartilage (cdECM), and heart (hdECM) tissue, PCL	Extrusion	(118)

is caused by horseradish peroxidase-catalyzed crosslinking with the consumption of hydrogen peroxide.

The 3D cell-laden hydrogels are fabricated by the sequential dropping of bioink containing polymers that are crosslinkable through the enzymatic reaction and hydrogen peroxide onto droplets of another bioink containing the polymer, horseradish peroxidase, and cells.

The 95% viability of enclosed mouse fibroblasts and subsequent elongation of the cells in a bioprinted hydrogel consisting of gelatin and hyaluronic acid derivatives suggest a high cytocompatibility of the here developed printing approach.

The existence of numerous polymers, including derivatives of polysaccharides, proteins, and synthetic polymers, crosslinkable through the horseradish peroxidase-catalyzed reaction, suggests that this approach holds great promise for the biofabrication of functional and structurally complex tissues (158).

5.3.2 *Hydrogel Bioinks*

For successful high-fidelity 3D hydrogel printing, it was found that two general conditions must be met: First, droplets must retain a raised 3D profile rather than spreading completely flat. However, liquid gel precursors printed onto an already gelled surface of the same kind typically exhibit complete spreading, as observed in several hydrogel systems. The second condition for successful 3D printing is that droplets on a surface must be sufficiently gelled so as not to coalesce with new droplets printed adjacent to or upon them (159).

5.3.2.1 *Thermoresponsive Polyisocyanide*

An approach to 3D deposition has been presented using a new class of fully synthetic, biocompatible poly(isocyanide) hydrogels that exhibit a reverse gelation temperature close to physiological conditions of 37°C (160). Being fully synthetic, poly(isocyanide) hydrogels are particularly attractive for tissue engineering, as their properties, such as hydrogel stiffness, polymer solubility, and gelation kinetics, can be precisely tailored according to process requirements. The synthesis of such a hydrogel runs as follows (160, 161):

Preparation 5–1: A solution of catalyst $Ni(ClO_4)_2 \times 6H_2O$ (1 mM) in toluene/ethanol (9:1) was added to a solution of 2-(2-(2-methoxyethoxy)-ethoxy)ethyl-(l)alaninyl-(*d*)-isocyanoalanine monomer in freshly distilled toluene (50 $mg\,ml^{-1}$ total concentration) with a catalyst: a monomer ratio of 1:4000 in an inert atmosphere. The reaction mixture was stirred at room temperature (20°C) for 96 h, and Fourier transform infrared (FTIR) analysis was performed over a range of 4000–400 cm^{-1} at a resolution of 4 cm^{-1}. After the FTIR results confirmed the completion of the reaction, the resulting polymer was precipitated three times from dichloromethane in diisopropyl ether and dried overnight in air. For printing, 50 mg of the poly(isocyanide) was dissolved in 10 ml of deionized water and was printed with a 3D-Bioplotter.

A gelatin methacrylate hydrogel was prepared as follows (162):

Preparation 5–2: A type A porcine skin gelatin (Sigma, Poznań, Warsaw) was dissolved at 10% w/v into phosphate buffered saline at 60°C and stirred until it was fully dissolved. A dose of 0.8 mL of methacrylic anhydride (Sigma) per gram of gelatin was added under constant stirring. After the reaction was completed, the mixture was dialyzed against distilled water using a 12–14 kD cut-off dialysis tubing (Spectra/Por, Rancho Dominguez, CA, USA) for 7 d at 40°C to remove the salts and the methacrylic acid. Then, the solution was lyophilized to generate a white porous foam and stored at −80°C until further use.

The feasibility of both 3D printing poly(isocyanide) hydrogels and of creating dual poly(isocyanide)-gelatin methacrylate hydrogel systems was demonstrated. Furthermore, the use of poly(isocyanide) as fugitive hydrogel to template structures within gelatin methacrylate hydrogels has been proposed.

This approach represents a robust and valid alternative to other commercial thermosensitive systems, such as those based on Pluronic F127, for the fabrication of 3D hydrogels through additive manufacturing technologies to be used as advanced platforms in tissue engineering (160).

5.3.2.2 Model of a Blood Vessel

Printing a raised and hollow alginate-based lumen as a simple model of a blood vessel has been achieved.

The preparation of hydrogel for printing was described (159). The components of the hydrogels are sodium alginate, agarose, high-strength porcine gelatin, sodium chloride, and calcium chloride.

5.3.2.3 *Tunable, Cell-Compatible Hydrogels*

A multimaterial bioink method using PEG crosslinking has been presented for expanding the biomaterial palette required for 3D bioprinting of more mimetic and customizable tissue and organ constructs (153).

Slightly crosslinked, soft hydrogels were produced from precursor solutions of various materials and 3D printed. The polymer solutions were slightly crosslinked with a long length of 5000 $g\ mol^{-1}$ using a chemical crosslinking agent, a homobifunctional PEG ending in two reactive groups. The polymers can be linear or branched, as well as have multifunctional groups for primary (bioink synthesis) and secondary (post-printing) crosslinking.

Since it is commercially available in many physical variants, such as linear, multiarm, molecular weight variation and also chemical variants, PEG is an ideal crosslinking agent (163,164).

Also, rheological and biological characterizations have been assessed, and the promise of this bioink synthesis strategy has been discussed.

5.3.3 *Dentin-Derived Hydrogel Bioink*

Recent studies in tissue engineering have adopted matrix-derived scaffolds as natural and cytocompatible microenvironments for tissue regeneration.

The dentin matrix, specifically, has been shown to be associated with a host of soluble and insoluble signaling molecules that can promote odontogenesis.

A novel bioink was developed by blending printable alginate (3% w/v) hydrogels with the soluble and insoluble fractions of the dentin matrix (165). The printing parameters were optimized along with the concentrations of the individual components of the bioink for print accuracy, cell viability and odontogenic potential.

While viscosity, and hence the printability of the bioinks, was greater in the formulations containing higher concentrations of alginate, a higher proportion of insoluble dentin matrix proteins significantly improved cell viability. A 1:1 ratio of alginate and dentin was found to be most suitable.

A high retention of the soluble dentin molecules was demonstrated within the 1:1 alginate dentin hydrogel blends, evidencing renewed interactions between these molecules and the dentin matrix post crosslinking. Moreover, at concentrations of 100 $\mu g\, ml^{-1}$, these soluble dentin molecules significantly enhanced odontogenic differentiation of stem cells from the apical papilla encapsulated in bioprinted hydrogels.

In summary, the proposed novel bioinks show a demonstrable cytocompatibility and natural odontogenic capacity, which can be used to reproducibly fabricate scaffolds with complex three-dimensional microarchitectures for regenerative dentistry (165).

5.3.4 *Decellularized Extracellular Matrix Materials*

Stem cell-laden decellularized extracellular matrix bioinks have been used for 3D printing of prevascularized and functional multimaterial structures (166). The printed structure composed of spatial patterning of dual stem cells improves cell-to-cell interactions and differentiation capability and promotes the functionality for tissue regeneration.

The so developed stem cell patch promoted a strong vascularization and tissue matrix formation *in vivo*. The patterned patch exhibited enhanced cardiac functions, reduced cardiac hypertrophy and fibrosis, increased migration from patch to the infarct area, neo-muscle and capillary formation along with improvements in cardiac functions (166). *In-vivo* studies were performed for investigating neovascularization and tissue formation, using animals such as mice and rats.

In addition, extrusion-based 3D printing strategies have the potential to fabricate 3D-bioprinted cardiac constructs by depositing cardiac cells with appropriate biomaterials (167).

Heart-derived decellularized extracellular matrices containing a complex mixture of various extracellular molecules provide a comprehensive microenvironmental niche similar to native cardiac tissue. However, a major concern is the insufficient vascularization and mimicking of the complex 3D architectural features, which can be tackled using 3D printing approaches (167).

The advantage and application of decellularized extracellular matrix-based hydrogels for the 3D printing of engineered cardiac tissues have been reviewed (167).

Also, the integration of electroactive materials within cardiac patches to improve the electrophysiological properties of the myocardium has been discussed (167).

5.3.4.1 *Cardiac Patches*

Biofabrication of cell supportive cardiac patches that can be directly implanted on myocardial infarct is a potential solution for myocardial infarction repair (168). Cardiac patches should be able to mimic myocardium extracellular matrix for rapid integration with the host tissue, raising the need to develop cardiac constructs with complex features. Cardiac patches should be electrically conductive, mechanically robust and elastic, biologically active and prevascularized.

The fabrication of a nanoreinforced hybrid cardiac patch, loaded with human coronary artery endothelial cells with improved electrical, mechanical, and biological behavior has been shown (168).

A safe UV exposure time with insignificant effect on cell viability was identified for methacrylated collagen micropatterning. The effects of carboxyl functionalized carbon nanotubes (CNTs) on the methacrylated collagen morphology and the alginate matrix morphology, the mechanical properties, electrical behavior, and cellular response were investigated at different CNT mass ratios.

An UV-integrated 3D-bioprinting technique was implemented to create hybrid hydrogel constructs consisting of CNT incorporated alginate framework and cell loaded methacrylated collagen. The compressive modulus, impedance, and swelling degree of hybrid constructs were assessed over 20 *d* of incubation in a culture medium at 37°C for different CNT mass ratios. The human coronary artery endothelial cell viability, proliferation, and differentiation in the context of the bioprinted hybrid constructs were assessed over 10 *d in vitro*. The functionalized CNTs provided a highly interconnected nanofibrous meshwork that significantly improved viscoelastic behavior and electrical conductivity of photo-crosslinked methacrylated collagen (168).

Alginate-coated CNTs provided a nanofilamentous network with fiber size of ~ 25 *nm* – 500 *nm*, which not only improve electrical and mechanical properties but also human coronary artery endothelial cell attachment and elongation compared to pristine alginate. The CNT reinforced 3D printed hybrid constructs presented significantly higher stiffness and electrical conductivity particularly in the physiologically relevant frequency range (~5 *Hz*).

The CNT-reinforced hybrid implants maintain a significantly higher swelling degree over 20 *d* of culturing compared to CNT-free hybrid constructs. For a selected CNT mass ratio, human coronary artery endothelial cells presented a significant cellular proliferation, migration, and differentiation (lumen-like formation) over 10 *d* of incubation *in vitro* (168).

5.3.5 *Silk-Based Bioink*

Silk fibroin offers a promising choice for bioink material (169). Nature has imparted several unique structural features in silk protein to ensure the spinnability by silkworms or spiders. The structure-property relationship has been modified by reverse engineering to further improve the shear-thinning behavior, high printability, cytocompatible gelation, and high structural fidelity.

In a recent review, it was attempted to summarize the advancements made in the field of 3D bioprinting in context of two major sources of silk fibroin: silkworm silk and spider silk, i.e., native and recombinant. The challenges faced by current approaches in processing silk bioinks, cellular signaling pathways modulated by silk chemistry and secondary conformations, gaps in knowledge, and future directions acquired for pushing the field further toward clinical applcations have been further elaborated (169).

Silk/PEG hydrogels were studied as self-standing bioinks for 3D printing in tissue engineering (170). The two components of the bioink, a silk fibroin protein and PEG, are both Food and Drug Administration approved materials in drug and medical device products. Mixing the PEG with silk induces a silk β-sheet structure formation and thus gelation and water insolubility due to physical crosslinking. A variety of constructs with high resolution, high shape fidelity, and homogeneous gel matrices were printed.

When human bone marrow mesenchymal stem cells are premixed with the silk solution prior to printing and the constructs are cultured in this medium, the cell-loaded constructs maintain their shape over at least 12 weeks.

The cells grow faster in the higher silk concentration (10% w/v) gel than in lower ones (e.g., 7.5% w/v and 5% w/v). This likely occurs due to the difference in material stiffness and the amount of residual PEG remaining in the gel related to material hydrophobicity.

The subcutaneous implantation of 7.5% w/v bioink gels with and without printed fibroblast cells in mice revealed that the cells survive and proliferate in the gel matrix for at least 6 weeks postimplantation (170).

5.3.6 *Nanoengineered Ionic-Covalent Entanglement Bioinks*

An enhanced nanoengineered ionic-covalent entanglement bioink for the fabrication of mechanically stiff and elastomeric 3D biostructures has been presented (171). Such bioink formulations combine nanocomposite and ionic-covalent entanglement strengthening mechanisms to print customizable cell-laden constructs for tissue engineering with high structural fidelity and mechanical stiffness.

A standard formulation of a nanoengineered ionic-covalent entanglement bioink consists of 1% κ-carrageenan, 10% gelatin methacrylate, and 2% nanosilicates (172).

Nanocomposite and ionic-covalent entanglement strengthening mechanisms complement each other through synergistic interactions, improving mechanical strength, elasticity, toughness, and flow properties beyond the sum of the effects of either reinforcement technique alone (171).

Herschel-Bulkley flow behavior shields encapsulated cells from excessive shear stresses during extrusion (171).

The Herschel-Bulkley fluid is a generalized model of a non-Newtonian fluid, in which the strain experienced by the fluid is related to the stress in a complicated, nonlinear way (173). This fluid model was introduced by Winslow Herschel and Ronald Bulkley in 1926 (174).

The encapsulated cells readily proliferate and maintain high cell viability over 120 *d* within the 3D printed structure, which is vital for long-term tissue regeneration (171). A unique aspect of the nanoengineered bioink is its ability to print much taller structures, with higher aspect ratios, than can be achieved with conventional bioinks without requiring secondary supports.

Thus, nanoengineered bioinks can be used to bioprint complex, large-scale, cell-laden constructs for tissue engineering with high structural fidelity and mechanical stiffness for applications in custom bioprinted scaffolds and tissue engineered implants (171).

5.3.7 *Living Skin Constructs*

A novel bioink made of gelatin methacrylamide and collagen doped with tyrosinase has been presented for the 3D bioprinting of living skin tissues (175).

Tyrosinase has the dual function of being an essential bioactive compound in the skin regeneration process and also as an enzyme to facilitate the crosslink of collagen and gelatin methacrylamide. Furthermore, enzyme crosslinking together with photocrosslinking can enhance the mechanical strength of the bioink. The experimental results indicated that the bioink is able to form stable 3D living constructs using the 3D bioprinting process. The cell culture shows three major cell lines (175):

1. Human melanocytes,
2. Human keratinocytes, and
3. Human dermal fibroblasts.

These exhibit high cell viabilities. The viability of these three cell lines is above 90%. The proliferation and scratching test show that tyrosinase can enhance the proliferation of human melanocytes, inhibit the growth and migration of human dermal fibroblasts and do not affect human keratinocytes significantly. Animal tests showed that the doped bioinks for 3D bioprinting can help form an epidermis and dermis, and thus have high potential as a skin bioink (175).

5.3.8 *Cell-Laden Scaffolds*

Methacrylated gelatin has been widely used as a tissue engineered scaffold material, but only low-concentration methacrylated gelatin hydrogels were found to be promising cell-laden bioinks with an excellent cell viability (176). A strategy was reported for the precise deposition of 5% w/v cell-laden methacrylated gelatin bioinks into controlled microarchitectures with high cell viability using extrusion-based 3D bioprinting. By adding gelatin into the methacrylated gelatin bioinks, a two-step crosslinking combining the rapid and reversible thermo-crosslinking of gelatin with irreversible photocrosslinking of methacrylated gelatin could be achieved (176).

The methacrylated gelatin/gelatin bioinks showed significant advantages in the processability, because the tunable rheology and the rapid thermo-crosslinking of the bioinks improved the shape fidelity after bioprinting.

The 5% w/v methacrylated gelatin with 8% w/v gelatin could be printed into 3D structures, which had the similar geometrical resolution as that of the structures printed by 30% w/v methacrylated gelatin bioinks. This printing strategy of methacrylated gelatin/gelatin bioinks may extensively extend the applications of methacrylated gelatin hydrogels for tissue engineering, organ printing, or drug delivery (176).

Also, another bioink was developed based on gellan gum-poly-(ethylene glycol) diacrylate double-network hydrogel (177). This bioink can deposit cell-laden hydrogels and biodegradable thermoplastics together. The biocompatibilities of hydrogels and the mechanical performances of thermoplastics can be merged to meet the necessary requirement of intervertebral disc regeneration.

The 3D cell-laden constructs were prepared using a multi-head bioprinting system. Two automatic switching printing heads which can dispense different materials simultaneously were used (177):

1. One loaded with cell-laden gellan gum-poly(ethylene glycol) diacrylate pre-gel mixture and maintained at 37°C. The hydrogels were extruded through a 25G flat tip needle.
2. The other printing head with a 400 μm nozzle contained a PLA wire and was set at 200°C in order to melt the polymer.

Computer-aided design models of various microstructures were used to fabricate different architectures (177). Subsequently, the scaffolds were achieved by a 5 *min* ultraviolet polymerization process at 365 *nm*, 50 $mW\,cm^{-2}$, and cultured under the same conditions as bone marrow stromal cells.

The viability and morphology of cells encapsulated in hydrogels were verified at 1 *d*, 3 *d*, and 7 *d* after printing. The viability tests were evaluated by staining using a live/dead cell assay kit. For the morphological study, each specimen was fixed in 4% paraformaldehyde, permeabilized by 0.1% Triton X-100, blocked with bovine serum albumin, and immune stained with rhodamine-conjugated phalloidin for F-actin filament and 4′,6-diamidino-2-phenylindole, cf. Figure 5.2, for nucleus in the dark. Then a confocal laser scanning microscopy was performed.

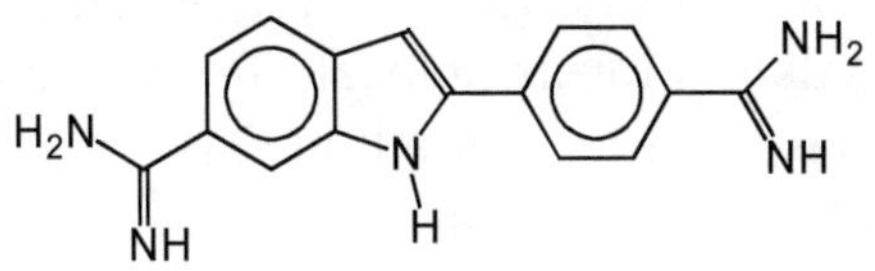

Figure 5.2 4′,6-Diamidino-2-phenylindole.

The bone marrow stromal cells encapsulated in the gellan gum-poly(ethylene glycol) diacrylate hydrogel were uniformly dispersed in the scaffold after the bioprinting process. They showed a high cell proliferation rate and remained viable above 90% during the culture time (177).

5.3.9 *Patient-Specific Bioinks*

The incorporation of biological factors has not been well explored in the past (178). As the importance of personalized medicine is becoming more clear, the need for the development of bioinks containing autologous/patient-specific biological factors for tissue engineering applications has become more evident.

Platelet-rich plasma is used as a patient-specific source of autologous growth factors that can be easily incorporated into hydrogels and printed into 3D constructs. A platelet-rich plasma contains a cocktail of growth factors enhancing angiogenesis, stem cell recruitment, and tissue regeneration.

The development of an alginate-based bioink that can be printed and crosslinked upon implantation through exposure to native calcium ions has been reported. Such a material can be used for the controlled release of platelet-rich plasma-associated growth factors which may ultimately enhance vascularization and stem cell migration (178).

5.4 Presurgical Simulation

A presurgical simulation method has been presented that uses interactive virtual simulation with 3D computer graphics data and microscopic observation of color-printed plaster models based on these data (24).

An interactive virtual simulation was performed by modifying the 3D data to imitate various surgical procedures, such as bone drilling, brain retraction, and tumor removal, with manipulation of a haptic device. Also, color-printed plaster models were produced using a selective laser sintering method (24).

5.5 Models with Integrated Soft Tactile Sensors

3D printed soft capacitive tactile sensors could be fabricated that respond to applied pressures in the form of a change in the device capacitance (179–181). The sensors consisted of a poly(acrylamide)-based ionic hydrogel (as electrodes) and a silicone-based dielectric elastomer.

5.6 Dental Applications

Computer-aided design (CAD) and computer-aided manufacturing (CAM) systems have been reviewed that can be used for dental restoration methods (182).

Dental research centers can create crowns, bridges, stone models, implants and different surgical, endodontic and orthodontic appliances with strategies that combine oral scanning, 3D printing, and CAD/CAM design. 3D printing has been utilized for the advancement of models. The advances in additive manufacturing and its developing applications in the field of dentistry have been reviewed (183).

It has been shown that temporary crowns can be 3D printed with adequate mechanical properties for intraoral use (184). In addition, the results of the study suggested that a 3D printable provisional restorative material allows for sufficient mechanical properties for intraoral use, despite the limited 3D printing accuracy of the low-cost stereolithography 3D printing system under investigation.

The technology of 3D printing has a particular resonance with dentistry, and with the advances in 3D imaging and modeling technologies such as cone beam computed tomography and intraoral scanning (185). With the relatively long history of the use of CAD CAM technologies in dentistry, it is becoming of increasing importance.

The use of 3D printing include the production of drill guides for dental implants, the production of physical models for prosthodontics, orthodontics and surgery, the manufacture of dental, craniomaxillofacial and orthopedic implants, and the fabrication of copings and frameworks for implant and dental restorations.

The types of the available 3D printing technologies and their various applications in dentistry and in maxillofacial surgery have been reviewed (185).

Medical imaging has been used to provide information for diagnostic and therapeutic purposes (186). The use of physical models provides added values in these applications. Rapid prototyping (RP) techniques have long been employed to build complex 3D models in medicine. The basics and applications of RP techniques in dentistry have been detailed (186):

1. Construction of a CAD model, including data acquisition, data processing, and the corresponding machines and CAD packages,
2. Typical RP systems and how to choose them, and
3. The potential use of RP techniques in dentistry.

Also, practical application examples of RP techniques in dentistry have been provided (186).

5.6.1 *Prosthetics*

Earlier, dental prosthesis, i.e., coping, crown, bridge, fixture, etc., fabrication greatly depended on the skills of dentists and the technicians (186,187). Then, RP techniques became an alternative method of fabrication. For example, patterns for dental crowns and implant structures can be fabricated using a RP machine (188). Dental crowns can be used to restore damaged or missing teeth.

5.6.1.1 Rapid Prototyping

In recent years, additive methods by employing RP have progressed rapidly in various fields of dentistry, as they have the potential to overcome known drawbacks of subtractive techniques such as fit problems (189).

Ever since the 1990s, RP techniques have been exploited to build complex 3D models in medicine. Recently, RP has proposed successful applications in various dental fields, such as fabrication of implant surgical guides, frameworks for fixed and removable partial dentures, wax patterns for the dental prosthesis, zirconia prosthesis and molds for metal castings, and maxillofacial prosthesis and, finally, complete dentures.

A comprehensive literature review of various RP methods has been presented, in particular in dentistry (189).

The various laboratory procedures employed in this method were assessed and it could be confirmed that the RP technique has been found to be substantially feasible in dentistry. With advancement in various RP systems, it is possible to benefit from this technique in different dental practices, particularly in implementing dental prostheses for different applications (189).

The chief benefits of RP techniques are the medical models that can be produced with undercuts, voids, intricate internal geometrical details and anatomical landmarks such as facial sinuses and neurovascular canals. The RP model is currently employed to improve medical diagnosis and to provide a precise surgical treatment plan. The technique would help to shorten surgery time and consequently reduce the risks for the patients.

5.6.1.2 Dental Crown

A dental crown model can be constructed from the inner surface to the outer surface. The inner geometrical data can be obtained either by scanning the surface of the tooth after tooth preparation or based on the profile of the implant, while the outer surface can be designed based on the scanned data from neighboring teeth and the teeth on the opposite side of the mouth and esthetic considerations. After the construction of the crown model, the model can be sliced and transferred to a RP machine to fabricate the crown pattern. The crown model is then investment cast to a metallic or ceramic crown. Through this process, the feedback on the design of the dental crown from the patient can be taken into consideration before the dental crown is fabricated (186).

5.7 Fluidic Devices

It is estimated that in the United States more than 16 million units per year of stored blood are required for transfusion purposes. There may be complications arising from the transfusion of blood components.

A fluidic device has been fabricated by a 3D printing technology for obtaining some analytical data. 3D printed fluidic devices enable quantitative evaluation of blood components in modified storage solutions for use in transfusion medicine (190).

5.8 3D Bioprinting of Tissues and Organs

The field of organ printing and the broader bioprinting field was first defined by Boland and colleagues in 2003 (191). Its feasibility was demonstrated in 2005 through the bioprinting of hamster ovarian cancer cells via an inkjet printer (192).

The 3D printing technologies for 3D scaffold engineering have been described in a monograph (193).

The cost-to-entry into the bioprinting field has decreased as more affordable bioprinters have become commercially available and open source 3D printers and other hobbyist 3D printers have been adapted to print biological materials and hydrogels (105).

The purpose of 3D bioprinting technology is to design and create functional 3D tissues or organs *in situ* for *in-vivo* applications (117). 3D cell printing, or additive biomanufacturing, allows the selection of biomaterials and cells (bioink), and the fabrication of cell-laden structures in high resolution. 3D cell-printed structures have also been used for applications such as research models, drug delivery and discovery, and toxicology.

Recently, numerous attempts have been made to fabricate tissues and organs by using various 3D printing techniques. The most commonly used 3D cell printing techniques with their advantages and drawbacks have been reviewed (117, 194–196).

3D bioprinting has been put forward with the technical progress in 3D printing and might be a possible way to solve the serious problem of human organ shortage in tissue engineering and regenerative medicine (197). Many research groups have been flung into this area and have already made some gratifying achievements. The background and development history of 3D bioprinting has been reviewed. Different approaches of 3D bioprinting are compared and the key factors of the printing process are illustrated. Also, existing challenges of 3D bioprinting were pointed out (197).

Bioprinting has little or no side effects in printed mammalian cells and can conveniently combine with gene transfection or drug delivery in the ejected living systems during the precise placement of tissue construction. With layer-by-layer assembly, 3D tissues with complex structures can be printed using scanned CT or MRI images. Vascular or nerve systems can be enabled simultaneously during the organ construction with digital control (198).

Recent advances have enabled the 3D printing of biocompatible materials, cells and supporting components in complex 3D functional living tissues (114, 199). 3D bioprinting is being applied in regenerative medicine to address the need for tissues and organs suitable for transplantation.

In comparison to non-biological printing, 3D bioprinting involves additional complexities, such as the choice of materials, cell types, growth and differentiation factors, and technical challenges related to the sensitivities of living cells and the construction of tissues.

These complexities require the integration of technologies from the fields of engineering, biomaterials science, cell biology, physics and medicine. Already 3D bioprinting has been used for the generation and transplantation of several tissues, including multilayered skin, bone, vascular grafts, tracheal splints, heart tissue, and cartilaginous structures. Other applications include the development of high-throughput 3D-bioprinted tissue models for research, drug discovery, and toxicology (114).

5.8.1 3D Bioprinting Techniques

The most recent advances in 3D bioprinting have been reviewed. The methods of extrusion, inkjet, stereolithography, and laser bioprinting have been detailed (139).

One of these three approaches is used in 3D bioprinting (117):

1. Biomimicry,
2. Autonomous self-assembly, and
3. Mini-tissues.

These methods allow the printed product to mature and function like the original tissue (111).

Biomimicry aims to produce a replica of the cellular and extracellular components of a tissue or an organ. Autonomous self-assembly has stem cells or embryonic cells as cellular components that develop into the desired architecture and function. Mini-tissues provide the basic building block of the organ that is used by the two strategies (1 and 2) mentioned above. The 3D structure is determined mostly by the image obtained through a 3D scanner, CT, MRI, and ultrasound imaging. The gathered image is then processed by CAD-based software to produce files, such as surface tesselation language or g-code, that can be read and processed by bioprinters (117).

A lot of 3D bioprinting methods are available. The specific techniques and materials that are used in 3D bioprinting have been detailed (117).

The issues of inkjet-based 3D bioprinting have been described in a monograph (200). Inkjet bioprinters can deliver a controlled amount of bioink to the desired printing surface, either by thermal or acoustic means, forcing the content to flow continuously or be expelled from the nozzle (117). Several commercial inkjet printers have been modified to print biomaterials rather than ink. A layer-by-layer construction of bioprinting can be achieved by accommodating an additional printing bed, which moves in the z-axis in a manner controllable in the micrometer range. Because of the availability of commercial products and ease of modification, inkjet bioprinters are often used in bioprinting of tissues and organs.

Thermal inkjet bioprinters contain electrical heating systems of 200°C–300°C in the printer heads. The heat creates bubbles, thus ejecting the bioink out of the orifice. It has been demonstrated that the exposure of living cells to such heat for a short period of time does not have a devastating effect on the viability of cells (198,201).

Acoustic inkjet bioprinters use piezoelectric materials to create pulses in the printer heads, thus breaking bioink into droplets, and forcing them out of a nozzle onto a biopaper. Piezoelectric crystals undergo a shape deformation when the applied voltage is changed. Thus, by supplying an alternating voltage at the desired frequency, droplet size and ejection directionality can be controlled. This proves to be advantageous for printing cells via thermal expulsion, as the cells are not damaged by heat and printing conditions, and controllability of the droplet size can be enhanced.

Laser-assisted bioprinting is based on the principle of laser-induced forward transfer. A high-energy laser pulse creates high-pressure bubbles in the thin biomaterial layer, ejecting it onto the desired place.

Extrusion bioprinting yields a continuous stream of bioink onto the designed stage through a pneumatic or mechanical extrusion system. Most of the commercial 3D printers used for non-biological purposes use this method, and definitely have a broader spectrum of printing material compared to bioprinters. Such bioprinters have the widest range of biomaterials, including hydrogels, biocompatible copolymers, and cell spheroids, which have viscosities that allow them to be printed.

Bio-electrospray and cell electrospinning bioprinting methods use an electric field between two charged electrodes to expel droplets or a continuous stream of liquid material (202). Here, a jet of liquid biomaterial is accelerated inside a needle by using a high direct current voltage between the charged needle and a ground electrode. The accelerated charged media creates a cone-like shape at the tip of needle, and forms either droplets or continuous fibers depending on the material property of liquid media. This method has been used to create scaffolds in tissue engineering without cellular components.

5.8.2 *Pigmented Human Skin Constructs*

A 3D bioprinting approach has been used to fabricate 3D pigmented human skin constructs (203). These constructs are obtained from using three different types of skin cells, i.e., keratinocytes, melanocytes and fibroblasts, from three different skin donors.

They exhibit a similar constitutive pigmentation (pale pigmentation) as the skin donors. A two-step drop-on-demand bioprinting strategy facilitated the deposition of cell droplets to emulate the epidermal melanin units, i.e., predefined patterning of keratinocytes and melanocytes at the desired positions, and the manipulation of the microenvironment to fabricate 3D biomimetic hierarchical porous structures found in native skin tissue.

The 3D bioprinted pigmented skin constructs were compared to the pigmented skin constructs that were fabricated by a conventional manual-casting approach. An in-depth characterization of both the 3D pigmented skin constructs indicated that they have a higher degree of resemblance to a native skin tissue with regards to the presence of well-developed stratified epidermal layers and the presence of a continuous layer of basement membrane proteins, as compared to the manually cast samples. The 3D bioprinting approach facilitates the development of 3D *in-vitro* pigmented human skin constructs for potential toxicology testing and fundamental cell biology research (203).

5.8.3 *Strategies for Tissue Engineering*

Over the past decades, many approaches have been developed to fabricate biomimetic extracellular matrices with the desired properties for engineering of functional tissues (204).

The 3D bioprinting technology has emerged as a possible solution by bringing unprecedented freedom and versatility in depositing biological materials and cells in a well-controlled manner in the 3D volumes, therefore achieving precision engineering of functional tissues. The application of 3D bioprinting to tissue engineering has been reviewed (204).

Here, the general strategies for printing functional tissue constructs are discussed. Then, different types of bioprinting with a focus on nozzle-based techniques and their respective advantages have been detailed. Also, the limitations of current technologies have been discussed (204).

5.8.3.1 Cell-Derived Decellularized Matrices

Cell-derived decellularized matrices are promising cell culture substrates for tissue engineering and regenerative medicine (205). However, they must be fabricated as desirably shaped 3D scaffolds for these applications because they do not retain macrostructures of tissue and organs.

In order to fabricate cell-derived decellularized matrices as 3D scaffolds, 3D template scaffolds have been employed. 3D printed scaffolds are available to fabricate cell-derived decellularized matrices as desirably shaped 3D scaffolds.

Cell-derived decellularized matrices were prepared on 3D printed PLA scaffolds (205). HT-1080 fibrosarcoma cells were seeded on the 3D printed PLA scaffolds and were cultured to deposit extracellular matrix components beneath the cells.

After the culture, the cells were removed from cell-scaffold constructs. In addition, deposited fibronectin was detected on the surface of 3D printed PLA scaffolds. These results indicate that the cell-derived decellularized matrices were successfully prepared as desirably shaped 3D scaffolds with the aid of 3D printed scaffolds. Also, the matrices exhibited cell adhesiveness.

These results indicate that 3D printing techniques are useful for their fabrication of cell-derived decellularized matrices as desirably shaped 3D scaffolds. Improved production of such desirably shaped 3D scaffolds of cell-derived decellularized matrices will expand the applications in tissue engineering and regenerative medicine (205).

5.8.4 Bone Tissue

Bone tissue is the most common type of hard tissue in the area of 3D bioprinting (117). The fabrication of bone tissue constructs with a mechanical strength comparable to native bone poses the key challenge in this area (115). A bone tissue construct needs to be interconnectively porous for vascularization and cell proliferation (2).

Various methods were developed in order to 3D bioprint such constructs. Hybrid constructs with strength-giving biopolymer using cell-laden solutions were fabricated to provide such structural integrity and strength, while others used a single cell-containing solution to bioprint bone tissue (115,206,207).

5.8.4.1 *Polymer Emulsion Coating*

Hydroxyapatite-based 3D scaffolding can be used for the fabrication of complex-shaped scaffolds to reconstruct bone defects (208). It was tried to improve the osteoinductivity and compressive strength of a hydroxyapatite-based 3D scaffold for bone regeneration.

Bone morphogenetic protein-2 loaded nanoparticles were prepared by a double emulsion-solvent evaporation method and incorporated onto the surface of 3D scaffolds using ε-poly(caprolactone) and nanoparticles emulsion solution.

The surface morphology of the scaffold was characterized using a scanning electron microscopy method. Its biocompatibility and osteogenic effects were evaluated *in vitro* using human mesenchymal stem cells. The *in-vivo* bone regeneration efficiency was determined using a rabbit calvarial bone defect model.

3D hydroxyapatite scaffolds with nanoparticles could be obtained using a ε-poly(caprolactone) coating process. Bone morphogenetic protein-2 loaded nanoparticles were uniformly distributed on the scaffold surface and bone morphogenetic protein-2 was gradually released.

The cumulative *in-vitro* release profile of morphogenetic protein-2 from the scaffold is shown in Figure 5.3.

The initial burst release profile in Figure 5.3 shows the slow release of BMP-2 of around 17% over the first 3 *d*, followed by its continual release for 30 *d*.

The nanoparticle coating improved the compressive strength of the scaffold. The cell proliferation, adhesion, and osteogenic differentiation properties were improved with ε-poly(caprolactone) bone morphogenetic protein-2 loaded nanoparticles coated scaffold. *In-vivo* experiments showed that the formation of new bone was significantly higher in the ε-poly(caprolactone) bone morphogenetic protein-2 loaded nanoparticles group than in the uncoated scaffold-implanted group.

So, the coating method using ε-poly(caprolactone) and nanoparticle emulsion solutions was useful not only to incorporate bone morphogenetic protein-2 loaded nanoparticles onto the surface of the

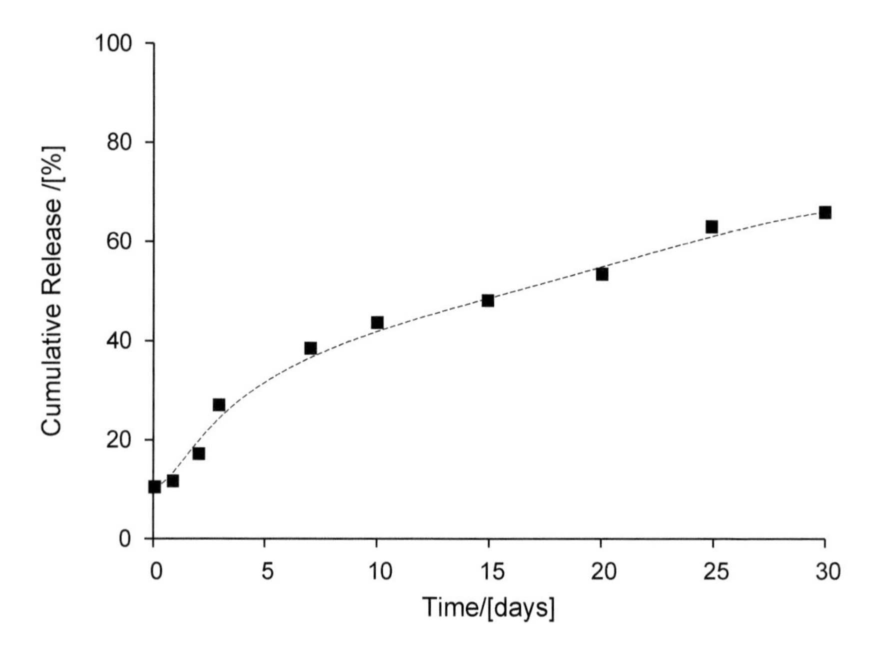

Figure 5.3 *In-vitro* cumulative release of morphogenetic protein-2 from scaffolds (208).

scaffold, but was also useful to improve the compressive strength, which enhanced the regeneration of the bone (208).

5.8.5 *Neuroregenerative Treatment*

Among several possible neuroregenerative treatment approaches that are known, 3D bioprinted scaffolds have the unique advantage of being highly modifiable, which promotes greater resemblance to the native biological architecture of *in-vivo* systems (209).

The high architectural similarity between printed constructs and *in-vivo* structures is believed to facilitate a greater capacity for repair of damaged nerve tissues.

The advances of several 3D bioprinting methods have been reviewed such as laser bioprinting, inkjet bioprinting, and extrusion-based printing (209). Also, the emergence of 4D printing has been discussed, which adds a dimension of transformation over time to traditional 3D printing.

5.8.6 *3D Tissues/Organs Combined with Microfluidics*

The regeneration of tissues and organs has benefited from various manufacturing technologies, especially 3D bioprinting (210). In 3D bioprinting, a cellular microenvironment, and biological functions of the native tissues/organs, cells and biomaterials are printed by a layer-by-layer assembly to form 3D biofunctional units.

However, there are still substantial differences between the 3D printed constructs and the actual tissues and organs, especially in microscale structures such as vascular networks. By manipulating controllable fluids carrying biomolecules, cells, organisms, or chemical agents, microfluidic techniques aim to integrate biological or chemical functional units into a chip. With its features of biocompatibility, flexible manipulation, and scale integration on the micro/nanoscale, microfluidics has been a tool that has enabled the generation of micro-tissues/organs with precise configurations.

There have been efforts to fabricate functional living tissues and artificial organs with complex structures via a combination of 3D bioprinting and the use of microfluidics.

The advances in microfluidics-assisted bioprinting in the engineering of tissues/organs have been discussed (210). In particular, examples of bioprinting with microfluidics, cell types, matrix materials, and target tissues/organs of the bioprinted 3D tissues/organs have been documented (210), cf. Table 5.5.

5.8.7 3D Microfibrous Constructs

Bioinks with shear-thinning/rapid solidification properties and strong mechanics are usually needed for the bioprinting of 3D cell-laden constructs (240). The generation of soft constructs from bioinks at low concentrations that are favorable for cellular activities are an important issue in this field.

A strategy for the fabrication of cell-laden constructs with tunable 3D microenvironments has been reported (240). This can be achieved by bioprinting of gelatin methacryloyl alginate core-sheath microfibers, where the alginate sheath serves as a template to support and confine the gelatin methacryloyl pre-hydrogel in the core during the extrusion process. Subsequent UV crosslinking can occur. This strategy minimizes the bioprinting requirements for the core bioink, and facilitates the fabrication of cell-laden gelatin methacryloyl constructs at low concentrations.

The capability of generating various alginate hollow microfibrous constructs was demonstrated using a coaxial nozzle setup. Also, the diffusibility and perfusability of the bioprinted hollow structures was shown. These properties are important for the tissue engineering applications.

The hollow alginate microfibers can be used as templates for generating cell-laden gelatin methacryloyl constructs with soft microenvironments by using gelatin methacryloyl pre-hydrogel as the bioink for the core phase during bioprinting. As such, gelatin methacryloyl constructs at extremely low concentrations of smaller than 2.0% could be extruded to effectively support the cellular activities, including proliferation and spreading for various cell types.

It is believed that this strategy provides broad opportunities in bioprinting of 3D constructs with cell-favorable microenvironments for applications in tissue engineering and pharmaceutical screening (240).

Table 5.5 Examples of bioprinted 3D tissues/organs combined with microfluidics (210).

Issue	Description
Method	Extrusion-based bioprinting with microfluidic modified printing nozzle
Cell types	Cartilage progenitor cells
Matrix materials	Alginate
Targets	Vascular tissues
References	(211)
Method	Extrusion-based bioprinting with microfluidic modified printing nozzle
Cell types	Human coronary artery smooth muscle cells
Matrix materials	CNT reinforced alginate
Targets	Vascular conduits
References	(212)
Method	Extrusion-based bioprinting with microfluidic modified printing nozzle
Cell types	L929 mouse fibroblasts
Matrix materials	Alginate
Targets	Lager-scale organs with built-in microchannels
References	(213)
Method	Extrusion-based bioprinting with microfluidic modified printing nozzle
Cell types	L929 mouse fibroblasts, smooth muscle cells, endothelial cells
Matrix materials	Alginate
Targets	Vascular circulation flow system
References	(214)
Method	Extrusion-based bioprinting with microfluidic modified printing nozzle
Cell types	HUVECs
Matrix materials	Alginate
Targets	Vascular network
References	(215)

Table 5.5 (cont.) Examples of bioprinted 3D tissues/organs combined with microfluidics (210).

Issue	Description
Method	Extrusion-based bioprinting with microfluidic modified printing nozzle
Cell types	Human embryonic kidney cells
Matrix materials	Alginate
Targets	Soft tissue scaffolds
References	(216)
Method	Extrusion-based bioprinting with microfluidic modified printing nozzle
Cell types	Muscle precursor cells
Matrix materials	PEG fibrinogen
Targets	Skeletal muscle tissue
References	(217)
Method	Extrusion-based bioprinting with cell printing in the receiving microfluidic plate
Cell types	HepG2
Matrix materials	Alginate
Targets	Liver
References	(218,219)
Method	Inkjet bioprinting with cell printing in the receiving microfluidic plate
Cell types	Hepatocytes and endothelial cells
Matrix materials	Fibronectin gelatin
Targets	Liver
References	(220)
Method	Inkjet bioprinting with cell printing in the receiving microfluidic plate
Cell types	Gelatin methacryloyl
Matrix materials	Fibronectin gelatin
Targets	Liver
References	(221)
Method	Extrusion-based bioprinting with cell printing in the receiving microfluidic plate
Cell types	HepG2 cells
Matrix materials	Alginate
Targets	Liver
References	(222)

Table 5.5 (cont.) Examples of bioprinted 3D tissues/organs combined with microfluidics (210).

Issue	Description
Method	Inkjet bioprinting with cell printing in the receiving microfluidic plate
Cell types	Hepatoma and glioma cells
Matrix materials	Alginate
Targets	Liver
References	(223)
Method	Sacrificial layer process and extrusion-based bioprinting of constructs with built-in microchannels
Cell types	Not mentioned in the original work
Matrix materials	Pluronic F127-diacrylate, (sacrificial material: Pluronic F127)
Targets	Microvascular networks
References	(224)
Method	Sacrificial layer process and extrusion-based bioprinting of constructs with built-in microchannels
Cell types	Endothelial cells, 10T1/2 cells, primary hepatocytes, stromal fibroblasts
Matrix materials	Agarose, alginate, PEG, fibrin, matrigel, (sacrificial material: carbohydrate glass)
Targets	Vascular tissues
References	(225)
Method	Sacrificial layer process and inkjet-based bioprinting of constructs with built-in microchannels
Cell types	HUVECs
Matrix materials	Collagen
Targets	Vascular tissues
References	(226)
Method	Sacrificial layer process and inkjet-based bioprinting of constructs with built-in microchannels
Cell types	HUVECs, normal human lung fibroblasts
Matrix materials	Collagen, fibrin, (sacrificial material: gelatin)
Targets	Vascular tissues
References	(227)

Table 5.5 (cont.) Examples of bioprinted 3D tissues/organs combined with microfluidics (210).

Issue	Description
Method	Sacrificial layer process and extrusion-based bioprinting of constructs with built-in micro-channels
Cell types	HepG2, NIH3T3, Mouse calvarial pre-osteoblast cells
Matrix materials	Gelatin methacryloyl, Star poly(ethylene glycol-*co*-lactide) acrylate, Poly(ethylene glycol) dimethacrylate, Poly(ethylene glycol) diacrylate, (sacrificial material: agarose)
Targets	Vascular tissues
References	(228, 229)
Method	Sacrificial layer process and extrusion-based bioprinting of constructs with built-in micro-channels
Cell types	HepG2/C3A cells, HUVECs
Matrix materials	Gelatin methacryloyl, (sacrificial material: agarose)
Targets	Vascularized liver tissue
References	(230)
Method	Sacrificial layer process and extrusion-based bioprinting of constructs with built-in micro-channels
Cell types	Human mesenchymal stem cells, Human neonatal dermal fibroblasts, HUVECs
Matrix materials	Fibrin, gelatin, (sacrificial material: Pluronic F127)
Targets	Thick vascularized tissues
References	(231)
Method	Sacrificial layer process and extrusion-based bioprinting of constructs with built-in micro-channels
Cell types	C3H/10T1/2, Human neonatal dermal fibroblasts, HUVECs
Matrix materials	Gelatin methacryloyl, (sacrificial material: Pluronic F127)
Targets	Vascularized tissues
References	(232)

Table 5.5 (cont.) Examples of bioprinted 3D tissues/organs combined with microfluidics (210).

Issue	Description
Method	Stereolithographic bioprinting of constructs with built-in microarchitecture
Cell types	HUVECs, C3H/10T1/2 cells, HepG2 cells
Matrix materials	Glycidal methacrylate-hyaluronic acid, Gelatin methacryloyl
Targets	Vascularized tissues
References	(233)
Method	Stereolithographic bioprinting of constructs with built-in microarchitecture
Cell types	hiPSCs-derived hepatic cells, HUVECs, ADSCs
Matrix materials	Glycidal methacrylate-hyaluronic acid, Gelatin methacryloyl
Targets	Vascularized hepatic constructs
References	(234)
Method	Block assembly and extrusion-based bioprinting of constructs with built-in microchannels
Cell types	HSFs, PASMCs
Matrix materials	Agarose (also as the sacrificial material)
Targets	Vascular tissues
References	(235)
Method	Sacrificial layer process and extrusion-based bioprinting of constructs with built-in microchannels
Cell types	Human immortalized PTEC cells
Matrix materials	Gelatin, fibrin, (sacrificial material: Pluronic F127)
Targets	Renal proximal tubules
References	(236)
Method	Block assembly and extrusion-based bioprinting of constructs with built-in microchannels
Cell types	Bone marrow stem cells, Schwann cells
Matrix materials	Agarose (also as the sacrificial material)
Targets	Nerve conduit
References	(237)

Table 5.5 (cont.) Examples of bioprinted 3D tissues/organs combined with microfluidics (210).

Issue	Description
Method	One-step fabrication of an organ-on-a-chip using cell/biomaterial printing
Cell types	HepG2 cells, HUVECs
Matrix materials	PCL, gelatin, collagen
Targets	Liver
References	(238)
Method	One-step fabrication of an organ-on-a-chip using cell/biomaterial printing
Cell types	Neonatal rat ventricular myocytes (NRVMs), human-induced pluripotent stem cell-derived cardiomyocytes (hiPS-CMs)
Matrix materials	Polydimethylsiloxane (PDMS)
Targets	Cardiac tissues
References	(239)

5.8.8 *Biosynthetic Cellulose Implants*

A biosynthetic cellulose fermentation technique for controlling 3D shape, thickness and architecture of the entangled cellulose nanofibril network has been developed (241). The resultant nanocellulose-based structures are useful as biomedical implants and devices. Furthermore, they are useful for tissue engineering and regenerative medicine, and for healthcare products.

The high water content of biosynthetic cellulose, around 99%, makes it attractive to be used as a hydrogel, which is known for its favorable biocompatible properties and low protein adsorption. Biosynthetic cellulose is a versatile material that can be manufactured in various sizes and shapes. The process of manufacturing is a biotechnological process and requires detailed control of bacterial proliferation, migration and production rate of cellulose.

Biosynthetic cellulose has been evaluated in several biomedical applications. It can be used for microsurgery (242) as vascular grafts (243), cartilage replacement (244), bone grafts (245), and meniscus implant (246).

A novel method has been developed to grow biosynthetic cellulose with a thickness which is theoretically unlimited and in any shape and robust structure.

A specially designed microfluidic medium administration system was used to gradually (e.g., continuously) increase the level of the culture media. The bacteria continue to produce a biosynthetic cellulose network when the media level is increased at the rate which matches the cellulose production. It was found that it is possible to optimize the thickness and strength of the resulting biosynthetic cellulose network by varying the rate and volume of media added.

A microfluidic administration system was developed in which media are added continuously at the air/bacteria interface with a minimum disturbance of bacteria (241). A porous mold system was used in which media diffuse gradually into the mold, which results in production of 3D biosynthetic cellulose structure of a predetermined shape and a robust structure. Also, a method was developed to introduce porosity by *printing* a porogen template with a controlled porosity which is inserted into 3D biosynthetic cellulose culture. This method enables the production of three-dimensional structures of biosynthetic cellulose with a controlled shape and porosity. Such materials have great potential for their application as implants and scaffolds for tissue engineering and regenerative medicine.

More particularly, embodiments of the present invention relate to systems and methods for the production and control of 3D architecture and morphology of nanocellulose biomaterials produced by bacteria using any biofabrication process, including the novel 3D bioprinting processes disclosed. Representative processes according to the invention involve control of the rate of production of biomaterial by bacteria achieved by meticulous control of the addition of fermentation media using a microfluidic system. In exemplary embodiments, the bacteria gradually grew up along the printed alginate structure that had been placed into the culture, incorporating it. After culture, the printed alginate structure was successfully removed, revealing porosity where the alginate had been placed. Porosity and interconnectivity of pores in the resultant 3D architecture can be achieved by porogen introduction using, e.g., inkjet printer technology.

Representative 3D bioprinting processes, systems, and devices used can employ one or more of the following (241):

1. Cells capable of synthesizing one or more extracellular biopolymers.
2. Culture media capable of maintaining cells under conditions conducive to the bioproduction of one or more extracellular polymers of interest and capable of administration by way of a microfluidic system.
3. Means for controlling the rate and volume of media administration.
4. A container or device for containing cells and administered media in a manner that allows for administration of additional media in a controlled manner.

The mechanical properties of biosynthetic cellulose structures are shown in Table 5.6.

Table 5.6 Mechanical properties of biosynthetic cellulose structures.

Material	Value/ [*MPa*]	Strain/ [%]	Modulus/ [*MPa*]
3D bioprinted cellulose	1.05	62	2.50
Static culture	0.39	57	0.80

5.8.9 Polysaccharides

Polysaccharides and their derivatives are highly attractive as biomaterials in the field of regenerative medicine, owing to their low cytotoxicity, hydrophilicity, and mechanical strength. In a chapter of a monograph, a short overview of 3D bioprinting, an emerging technology in regenerative medicine, has been given (247).

In addition, after a general introduction on polysaccharides and their derivatives the use of bioinks for printing of various types of structures, scaffolds and their characterization has been covered. Also, the application of bioprinted scaffolds in tissue engineering, i.e., cartilage, bone, and skin, is discussed in detail (247).

5.8.10 *Corneal Transplants*

The cornea is the transparent anterior part of the eye, which is essential for seeing. Corneal blindness affects millions of people worldwide.

Traditional corneal transplants from deceased donors have only a poor long-term success rate due to the lack of epithelial renewal (248,249).

The delivery of *in-vitro* expanded autologous cells to the corneal surface has been introduced as a possible treatment for patients suffering from unilateral or partially bilateral limbal stem cell deficiency (250,251).

Corneal transplantation constitutes one of the leading treatments for severe cases of loss of corneal function (252). Due to its limitations, a concerted effort has been made by tissue engineers to produce functional, synthetic corneal prostheses as an alternative recourse.

Bioinks have been developed for bioprinting of 3D corneal structures (253). Recombinant human laminin and human-sourced collagen I served as the bases for these functional bioinks.

Two established laser-assisted bioprinting setups were used. These were based on laser-induced forward transfer, with different laser wavelengths and appropriate absorption layers.

Three types of corneal structures were bioprinted:

1. Stratified corneal epithelium using human embryonic stem cell-derived limbal epithelial stem cells,
2. Lamellar corneal stroma using alternating acellular layers of bioink and layers with human adipose tissue-derived stem cells, and
3. Structures with both a stromal and epithelial part.

The printed constructs were evaluated for their microstructure, cell viability and proliferation, and key protein expression. The 3D printed stromal constructs were also implanted into porcine corneal organ cultures.

Initially, the laser-printed human embryonic stem cell-derived limbal epithelial stem cells were spherical in morphology, but recovered their normal polygonal morphology during culture.

It was concluded that 3D bioprinting is a promising method for the efficient fabrication of layered cornea-mimicking structures. Laser-assisted bioprinting with laser-induced forward transfer offers advantages over many other 3D bioprinting technologies, such as printing high-resolution 3D structures from viscous bioinks (253, 254).

In addition, corneal structures have been fabricated that resemble the structure of the native human corneal stroma using an existing 3D digital human corneal model and a suitable support structure. These were 3D bioprinted from an in-house collagen-based bioink containing encapsulated corneal keratocytes. Keratocytes exhibited high cell viability both at day 1 post-printing of greater than 90%, and at day 7 83% (252).

5.8.11 *Hydrogels from Collagen*

A single-step drop-on-demand bioprinting strategy has been proposed to fabricate hierarchical porous collagen-based hydrogels (255). Printable macromolecule-based bioinks using poly(*N*-vinyl-2-pyrrolidone) have been developed and printed in a drop-on-demand manner to manipulate the porosity within the multilayered collagen-based hydrogels by altering the collagen fibrillogenesis process.

The experimental results indicated that hierarchical porous collagen structures can be achieved by controlling the number of macromolecule-based bioink droplets printed on each printed collagen layer. This facile single-step bioprinting process could be useful for the structural design of collagen-based hydrogels for various tissue engineering applications (255).

5.8.12 *Dissolved Cellulose*

Dissolved cellulose has been 3D bioprinted to produce complex structures with ordered interconnected pores (256). This process consists of the dissolution of dissolving pulps in *N*-methylmorpholine-*N*-oxide (NMMO), multilayered dispensing, water removal of NMMO and freeze-drying.

An aqueous solution of NMMO can physically dissolve cellulose without any pretreatment (257).

The 3D bioprinting of a cellulose/NMMO solution at 70°C was analogous to that of thermoplastics by the process of melting and solidification to produce cellulose/NMMO objects in the solid form (256).

However, 3D bioprinting of cellulose/NMMO solution at a higher temperature than 70°C could produce cellulose/NMMO objects in a gel form. As such, a small amount of water was added dropwise to the top surface of the gel via a dropper at intervals of every 5 layers printed in order to precipitate cellulose fibers quickly and keep the printed structure from collapsing.

The cellulose could be regenerated by the usage of water. Afterwards, a freeze-drying treatment maintained the 3D bioprinted structures without collapsing. The final cellulose object produced in the solid form had a higher complex degree of inner structures than the one from the gel form, e.g., a controlled smaller pore size, thinner fibers and interconnected pores on the side wall.

The final cellulose products showed a remarkable Young's compressive modulus of 12.9 *MPa* and a tensile modulus of 160.6 *MPa*, cf. Figures 5.4 and 5.5. So, the feasibility of cellulose was demonstrated as an alternative method to produce complex structures (256).

5.8.13 *Hydrogels from Hyaluronic Acid and Methyl cellulose*

Hydrogels have been used in a wide range of biomedical applications and show great potential in the field of tissue engineering (258). Hydrogels are 3D crosslinked scaffolds of water-soluble polymers, which form a macromolecular network capable of retaining high water content. Due to their often poor mechanical integrity, hydrogels are classified as soft gels with structural similarity to some human soft tissues (259). However, their hydrophilic polymer networks enable the diffusion of glucose and other nutrients, thus supporting the growth of cells. Additionally, by altering the concentration of hydrogel components, mechanical properties can often be tailored (258).

Hydrogels containing hyaluronic acid and methyl cellulose have shown promising results for 3D bioprinting applications (258). However, several parameters influence the applicability of bioprinting and there were only a few data in the literature characterizing

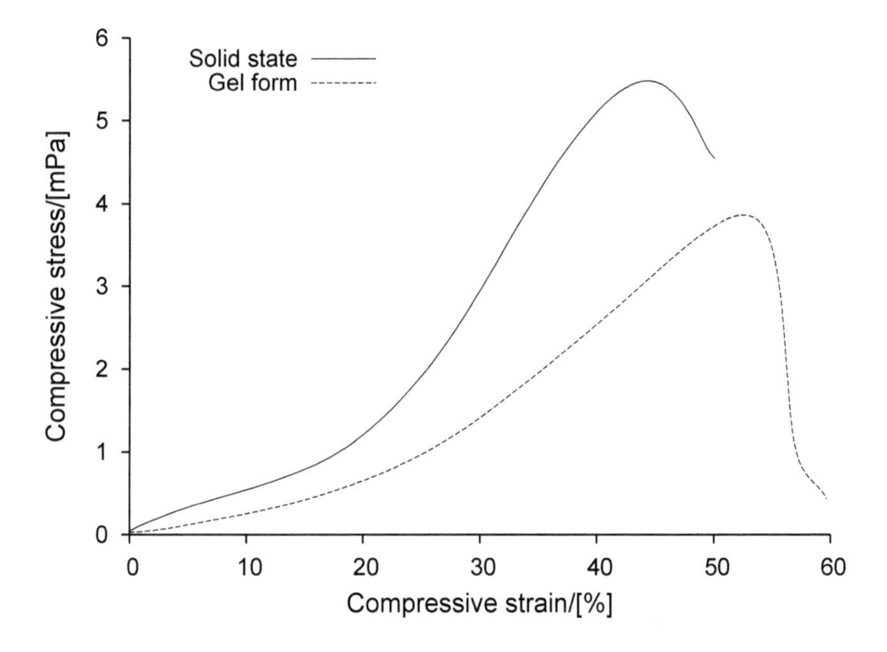

Figure 5.4 Compressive stress-strain curves for the 3D printed cellulose products (15 mm thick) (256).

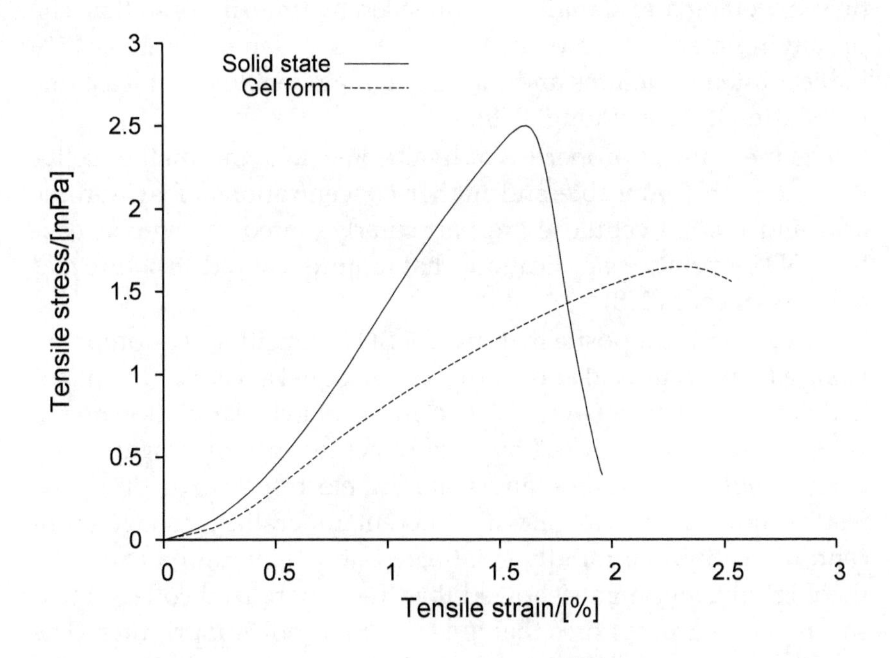

Figure 5.5 Tensile stress-strain curves for the 3D printed cellulose scaffolds (5 mm thick) (256).

hyaluronic acid and methyl cellulose. Several compositions were assessed for printability, swelling and stability over time, rheological and structural behavior, and viability of mesenchymal stem cells.

Hyaluronic acid and methyl cellulose blends behave as viscous solutions at 4°C and have faster gelation times at higher temperatures, typically gelling upon reaching 37°C. It was found that the storage, loss and compressive moduli are dependent on the concentration of hyaluronic acid and methyl cellulose and the incubation time at 37°C. Also, the compressive modulus is strain-rate-dependent. Swelling and stability is influenced by time, more so than the pH environment. The mesenchymal stem cell viability is above 75% in bioprinted structures and the cells remain viable for at least one week after 3D bioprinting (258).

The mechanical properties of hyaluronic acid and methyl cellulose are highly tuneable and higher concentrations of hyaluronic acid and methyl cellulose are particularly suited to cell-encapsulated 3D bioprinting applications that require scaffold structure and delivery of cells (258).

A gel-based composite material for 3D bioprinting was prepared using a hyaluronic acid and methyl cellulose gel as carrier and nano-hydroxyapatite/collagen (260). The effects of gel formulation on the gel process were analyzed by *in-vitro* gel testing, rheological analysis, *in-vitro* degradation and scanning electron microscopy. The results showed that this gel with different mineralized collagen concentration could be rapidly gelatinized at 37°C within 3 *min*. The rheological experiments showed that the mineralized collagen had shear-thinning properties that are suitable for 3D bioprinting. The results also showed that the degradation rate was similar in the first 10 *d*, but the degradation rate was slower with the increase of mineralized collagen concentration. The porous structure and high porosity of the hydrogel were analyzed by scanning electron microscopy (SEM). It was found that the porous structure of the crosslinked sample was better (260).

5.8.14 *Stem Cells*

Stem cells, such as human-induced pluripotent stem cells, have driven a paradigm shift in tissue regeneration and the modeling of human disease, and represent an unlimited cell source for tissue regeneration and the study of human disease (261). The ability to reprogram patient-specific cells may yield an enhanced understanding of some disease mechanisms and phenotypic variability.

3D bioprinting has been successfully performed using multiple stem cell types of different lineages and potency. The type of 3D bioprinting employed ranged from microextrusion bioprinting, inkjet bioprinting, laser-assisted bioprinting, to newer technologies such as scaffold-free spheroid-based bioprinting. The recent advances, applications, limitations and future of 3D bioprinting using stem cells by organ systems have been reviewed (261).

5.8.15 Autografts

Autografts are the current gold standard for large peripheral nerve defects in clinics despite the frequently occurring side effects like donor site morbidity (262). Hollow nerve guidance conduits have been proposed as alternatives to autografts, but fail to bridge gaps exceeding 3 *cm* in humans.

Internal nerve guidance conduits guidance cues like microfibers are believed to enhance hollow nerve guidance conduits by giving additional physical support for directed regeneration of Schwann cells and axons.

A new 3D *in-vitro* model has been reported that allows the evaluation of different intraluminal fiber scaffolds inside a complete nerve guidance conduit. The performance of electrospun PCL microfibers inside 5 *mm* long PEG conduits was investigated in neuronal cell and dorsal root ganglion cultures *in vitro*.

Z-stack confocal microscopy revealed the aligned orientation of neuronal cells along the fibers throughout the whole nerve guidance conduit length and depth (262). The number of living cells in the center of the scaffold was not significantly different than the tissue culture plastic control.

For *ex-vivo* analysis, dorsal root ganglions were placed on top of fiber-filled nerve guidance conduits to simulate the proximal nerve stump. In 21 *d* of culture, Schwann cells and axons infiltrated the conduits along the microfibers with 2.2 ± 0.37 *mm* and 2.1 ± 0.33 *mm*, respectively. It was concluded that this *in-vitro* model can help define internal nerve guidance conduit scaffolds in the future by comparing different fiber materials, composites and dimensions in one setup prior to animal testing (262).

5.8.16 Drug-Eluting Coronary Stents

Coronary artery disease has endangered people's lives in modern society, and is one of the common cardiovascular diseases (263). It is a highly prevalent disease with high mortality, readmission and complication rates. For treating cardiovascular diseases, the treatment of coronary artery disease is very important, with coronary intervention being the primary method of treatment. So far, the development of coronary intervention involves three stages (263):

1. Percutaneous transluminal coronary angioplasty,
2. Implanting bare-metal stents, and
3. Implanting drug-eluting stents.

Although the metal composition of a stent body, polymer carrier and antiproliferative drug have been much improved, two dilemmas remain unsolved: The metal material of stent and its cylindrical structure. The problems are as follows:

The permanent existence of metallic stents could bring about complications like late in-stent thrombosis, chronic inflammation, restenosis and stent fracture. Meanwhile, patients have to take life-long anti-platelet drugs which increase the risk of bleeding.

The diameter of a coronary artery gradually becomes smaller and sometimes the gap between the diameters of the proximal and distal segments of culprit arteries could be remarkable, which makes the lumen of vessels sharply tapered. These days, a traditional cylindrical-shaped stent may not be suitable for these cone-shaped arteries either by using a single stent or two stent technique, because implanting one stent may not expand efficiently due to the mismatched stent diameter and vessel diameter, thus leading to in-stent thrombosis, incomplete stent apposition, coronary dissection or in-stent restenosis.

Implanting two stents, on the other hand, could increase the risk of restenosis and thrombosis due to the overlapping area between stents, raising the total cost. Additionally, a traditional cylindrical stent cannot match the heteromorphic coronary artery such as an aneurysm.

A method has been presented using 3D printing technology to produce biomimetic drug-eluting coronary stents, which consists of the following steps (263):

1. According to coronary angiography, measuring the diameter of the diseased coronary artery by quantified coronary angiography (QCA), and designing the coronary stent suitable for the patient by three-dimensional reconstruction,
2. Utilizing fused deposition modeling/manufacturing (FDM), melted extrusion modeling (MEM), selective laser sintering (SLS), selective laser melting (SLM) or other additive manufacturing techniques to create a 3D printing platform, and
3. Melt extrusion nozzles, jet nozzles, SLS or SLM are used to manage biodegradable materials on the 3D printing platform of a coronary stent: High-energy beam selected melting or sintering material, or other materials, are controlled with hierarchical algorithm and molding control software to produce the biodegradable drug-eluting coronary stents.

Preferably, the biodegradable material is poly(*L*-lactic acid) (PLLA). The antiproliferative drug and poly(*L*-lactide-*co*-*D*,*L*-lactide (PDLLA) (264) are mixed at a ratio of 1:1 (263).

The biodegradable materials in biomimetic drug-eluting coronary stent produced by 3D printing technology are the same as that of a conventional stent made by PLLA material in hollow structure. The only difference is that the former can be personalized according to the patient's vessel (263).

5.9 Biomedical Devices

Applications for 3D biomedical devices are the restoration of 3D anatomic defects and the reconstruction of complex organs with intricate 3D microarchitecture, e.g., liver, lymphoid organs (265). Other examples are scaffolds for stem cell differentiation. Another example of such a need are anatomic defects in the craniomaxillofacial complex caused by cancer, trauma, and congenital defects. The proper restoration of these defects requires functional nerves, vessels, muscles, ligaments, cartilage, bone, lymph nodes and glands.

Various approaches based on tissue engineering principles have been explored to regenerate other functional tissues that are relevant to maxillofacial tissue regeneration. In tissue engineering, scaffolds are critical to providing the structure for cell infiltration and proliferation, space for extracellular matrix generation and remodeling, biochemical cues to direct cell behavior, and physical connections for injured tissue.

When constructing such scaffolds, the design of the architecture on the macro, micro, and nano level is important for structural, nutrient transport, and cell-matrix interaction conditions (266–268).

The macroarchitecture is the overall shape of the device, which can be complex; for example, the patient and organ specificity, and anatomical features. The microarchitecture reflects the tissue architecture, such as pore size, shape, porosity, spatial distribution, and pore interconnection. The nanoarchitecture is surface modification, e.g., biomolecule attachment for cell adhesion, proliferation, and differentiation.

The importance of diffusion in scaffolds and how it is influenced by porosity, permeability, architecture, and nutrient mixing has been emphasized (266). Also, methods for measuring porosity and permeability have been discussed.

The recent advances in both computational topology design and solid free-form fabrication allow the creation of scaffolds with a controlled architecture (267).

Cells are inherently sensitive to local mesoscale, microscale, and nanoscale patterns of chemistry and topography (268). The approaches to control cell behavior through the nanoscale engineering of material surfaces have been reviewed. The applications were found to have far-reaching implications, including medical implants, cell supports, and materials that can be used as instructive three-dimensional environments for tissue regeneration.

5.10 Soft Somatosensitive Actuators

Humans possess manual dexterity, motor skills, and other physical abilities that rely on feedback provided by the somatosensory system (269).

A method has been reported for creating soft somatosensitive actuators using embedded 3D printing, which are innervated with multiple conductive features that simultaneously enable haptic, proprioceptive, and thermoceptive sensing.

This manufacturing approach enables the seamless integration of multiple ionically conductive and fluidic features within elastomeric matrices to produce soft somatosensitive actuators with the desired bioinspired sensing and actuation capabilities. Each printed sensor is composed of an ionically conductive gel that exhibits both long-term stability and a hysteresis-free performance.

Multiple soft somatosensitive actuators can be combined into a soft robotic gripper that provides proprioceptive and haptic feedback via embedded curvature, inflation, and contact sensors, including deep and fine touch contact sensors. The multimaterial manufacturing platform enables complex sensing motifs to be easily integrated into soft actuating systems (269).

References

1. B.C. Gross, J.L. Erkal, S.Y. Lockwood, C. Chen, and D.M. Spence, *Analytical Chemistry*, Vol. 86, p. 3240, April 2014.
2. S. Bose, S. Vahabzadeh, and A. Bandyopadhyay, *Materials Today*, Vol. 16, p. 496, 2013.
3. L. Zhang, Y. Morsi, Y. Wang, Y. Li, and S. Ramakrishna, *Japanese Dental Science Review*, Vol. 49, p. 14, 2013.
4. F. Rengier, A. Mehndiratta, H. von Tengg-Kobligk, C.M. Zechmann, R. Unterhinninghofen, H.-U. Kauczor, and F.L. Giesel, *International Journal of Computer Assisted Radiology and Surgery*, Vol. 5, p. 335, July 2010.
5. M. Mahadevappa, *Radiographics*, Vol. 22, p. 949, 2002.
6. H. Von Tengg-Kobligk, T.F. Weber, F. Rengier, D. Kotelis, P. Geisbüsch, D. Böckler, H. Schumacher, and S. Ley, *Journal of Cardiovascular Surgery*, Vol. 49, p. 429, 2008.
7. F. Rengier, T.F. Weber, F.L. Giesel, D. Böckler, H.-U. Kauczor, and H. von Tengg-Kobligk, *American Journal of Roentgenology*, Vol. 192, p. W255, 2009.
8. D.H. Frakes, M.J.T. Smith, J. Parks, S. Sharma, M. Fogel, and A.P. Yoganathan, *Journal of Cardiovascular Magnetic Resonance*, Vol. 7, p. 425, 2005.
9. H.K. Hahn, W.S. Millar, O. Klinghammer, M.S. Durkin, P.K. Tulipano, and H.-O. Peitgen, *Methods Archive*, Vol. 43, p. 376, 2004.

10. T. Boland, T. Xu, B. Damon, and X. Cui, *Biotechnology Journal*, Vol. 1, p. 910, September 2006.
11. P.G. Campbell and L.E. Weiss, *Expert Opinion on Biological Therapy*, Vol. 7, p. 1123, 2007.
12. W.L. Ng, J.M. Lee, W.Y. Yeong, and M.W. Naing, *Biomaterials Science*, Vol. 5, p. 632, 2017.
13. K. Qiu, G. Haghiashtiani, and M.C. McAlpine, *Annual Review of Analytical Chemistry*, Vol. 11, p. in press, 2018.
14. G. Forgacs, F.S. Marga, and C. Norotte, Self-assembling multicellular bodies and methods of producing a three-dimensional biological structure using the same, US Patent 8 143 055, assigned to The Curators of the University of Missouri (Columbia, MO), March 27, 2012.
15. F. Liu, C. Liu, Q. Chen, Q. Ao, X. Tian, J. Fan, H. Tong, and X. Wang, *International Journal of Bioprinting*, Vol. 4, p. 1, 2018.
16. M. Vukicevic, B. Mosadegh, J.K. Min, and S.H. Little, *JACC: Cardiovascular Imaging*, Vol. 10, p. 171, 2017.
17. G. Wurm, B. Tomancok, P. Pogady, K. Holl, and J. Trenkler, *Journal of Neurosurgery*, Vol. 100, p. 139, 2004.
18. K. Qiu, Z. Zhao, G. Haghiashtiani, S.-Z. Guo, M. He, R. Su, Z. Zhu, D.B. Bhuiyan, P. Murugan, F. Meng, S.H. Park, C.-C. Chu, B.M. Ogle, D.A. Saltzman, B.R. Konety, R.M. Sweet, and M.C. McAlpine, *Advanced Materials Technologies*, Vol. 3, p. 1700235, 2018.
19. N. Wake, H. Chandarana, W.C. Huang, S.S. Taneja, and A.B. Rosenkrantz, *Clinical Radiology*, Vol. 71, p. 610, 2016.
20. M. Kusaka, M. Sugimoto, N. Fukami, H. Sasaki, M. Takenaka, T. Anraku, T. Ito, T. Kenmochi, R. Shiroki, and K. Hoshinaga, Initial experience with a tailor-made simulation and navigation program using a 3-D printer model of kidney transplantation surgery, in *Transplantation Proceedings*, Vol. 47, pp. 596–599. Elsevier, 2015.
21. J.R. Anderson, W.L. Thompson, A.K. Alkattan, O. Diaz, R. Klucznik, Y.J. Zhang, G.W. Britz, R.G. Grossman, and C. Karmonik, *Journal of NeuroInterventional Surgery*, Vol. 8, 2015.
22. D. Schmauss, S. Haeberle, C. Hagl, and R. Sodian, *European Journal of Cardio-Thoracic Surgery*, Vol. 47, p. 1044, 2015.
23. S. Mottl-Link, M. Hübler, T. Kühne, U. Rietdorf, J.J. Krueger, B. Schnackenburg, R. De Simone, F. Berger, A. Juraszek, and H.-P. Meinzer, *The Annals of Thoracic Surgery*, Vol. 86, p. 273, 2008.
24. M. Oishi, M. Fukuda, N. Yajima, K. Yoshida, M. Takahashi, T. Hiraishi, T. Takao, A. Saito, and Y. Fujii, *Journal of Neurosurgery*, Vol. 119, p. 94, July 2013.
25. K. Kondo, N. Harada, H. Masuda, N. Sugo, S. Terazono, S. Okonogi, Y. Sakaeyama, Y. Fuchinoue, S. Ando, D. Fukushima, J. Nomoto, and M. Nemoto, *Acta Neurochirurgica*, Vol. 158, p. 1213, Jun 2016.

26. C.M. Smith, T.D. Roy, A. Bhalkikar, B. Li, J.J. Hickman, and K.H. Church, *Tissue Engineering Part A*, Vol. 16, p. 717, 2010.
27. G.-S. Lee, J.-H. Park, U.S. Shin, and H.-W. Kim, *Acta Biomaterialia*, Vol. 7, p. 3178, 2011.
28. Y. Yan, F. Cui, R. Zhang, and Y. Hu, *Mater. Rev*, Vol. 14, p. 11, 2000.
29. A.E. Jakus, A.L. Rutz, S.W. Jordan, A. Kannan, S.M. Mitchell, C. Yun, K.D. Koube, S.C. Yoo, H.E. Whiteley, C.-P. Richter, R.D. Galiano, W.K. Hsu, S.R. Stock, E.L. Hsu, and R.N. Shah, *Science Translational Medicine*, Vol. 8, p. 358ra127, 2016.
30. A. Aied, W. Song, W. Wang, A. Baki, and A. Sigen, *Bioprinting*, 2018.
31. D. Han, Z. Lu, S.A. Chester, and H. Lee, *Scientific Reports*, Vol. 8, p. 1963, 2018.
32. C. Sun, N. Fang, D. Wu, and X. Zhang, *Sensors and Actuators A: Physical*, Vol. 121, p. 113, 2005.
33. X. Zheng, J. Deotte, M.P. Alonso, G.R. Farquar, T.H. Weisgraber, S. Gemberling, H. Lee, N. Fang, and C.M. Spadaccini, *Review of Scientific Instruments*, Vol. 83, p. 125001, 2012.
34. H. van der Linden, W. Olthuis, and P. Bergveld, *Lab Chip*, Vol. 4, p. 619, 2004.
35. H. Yusuke and Y. Ryo, *Macromolecular Chemistry and Physics*, Vol. 210, p. 2160, 2008.
36. W. Li, B. Belmont, J.M. Greve, A.B. Manders, B.C. Downey, X. Zhang, Z. Xu, D. Guo, and A. Shih, *Medical Physics*, Vol. 43, p. 5577, 2016.
37. M. Lazebnik, E.L. Madsen, G.R. Frank, and S.C. Hagness, *Physics in Medicine & Biology*, Vol. 50, p. 4245, 2005.
38. M.O. Culjat, D. Goldenberg, P. Tewari, and R.S. Singh, *Ultrasound in Medicine and Biology*, Vol. 36, p. 861, June 2010.
39. A.I. Farrer, H. Odéen, J. de Bever, B. Coats, D.L. Parker, A. Payne, and D.A. Christensen, *Journal of Therapeutic Ultrasound*, Vol. 3, p. 9, 2015.
40. M. Belohlavek, V.B. Bartleson, and M.E. Zobitz, *Echocardiography*, Vol. 18, p. 565, October 2001.
41. S. Langeland, J. D'hooge, T. Claessens, P. Claus, P. Verdonck, P. Suetens, G.R. Sutherland, and B. Bijnens, *IEEE Transactions on Ultrasonics, Ferroelectrics, and Frequency Control*, Vol. 51, p. 1537, November 2004.
42. M. McDonald, S. Lochhead, R. Chopra, and M.J. Bronskill, *Physics in Medicine & Biology*, Vol. 49, p. 2767, 2004.
43. Y. Yuan, C. Wyatt, P. Maccarini, P. Stauffer, O. Craciunescu, J. MacFall, M. Dewhirst, and S.K. Das, *Physics in Medicine & Biology*, Vol. 57, p. 2021, 2012.

44. R. Öpik, A. Hunt, A. Ristolainen, P.M. Aubin, and M. Kruusmaa, Development of high fidelity liver and kidney phantom organs for use with robotic surgical systems, in *2012 4th IEEE RAS EMBS International Conference on Biomedical Robotics and Biomechatronics (BioRob)*, pp. 425–430, June 2012.
45. F. Adams, T. Qiu, A. Mark, B. Fritz, L. Kramer, D. Schlager, U. Wetterauer, A. Miernik, and P. Fischer, *Annals of Biomedical Engineering*, Vol. 45, p. 963, Apr 2017.
46. K. Knox, C.W. Kerber, S.A. Singel, M.J. Bailey, and S.G. Imbesi, *American Journal of Neuroradiology*, Vol. 26, p. 1428, 2005.
47. L. Allard, G. Soulez, B. Chayer, Z. Qin, D. Roy, and G. Cloutier, *Medical Physics*, Vol. 40, p. 063701, 2013.
48. A.E. Forte, S. Galvan, F. Manieri, F.R. y Baena, and D. Dini, *Materials & Design*, Vol. 112, p. 227, 2016.
49. C.C. Ploch, C.S.S.A. Mansi, J. Jayamohan, and E. Kuhl, *World Neurosurgery*, Vol. 90, p. 668, June 2016.
50. N. Kadoya, Y. Miyasaka, Y. Nakajima, Y. Kuroda, K. Ito, M. Chiba, K. Sato, S. Dobashi, T. Yamamoto, N. Takahashi, M. Kubozono, K. Takeda, and K. Jingu, *Medical Physics*, Vol. 44, p. 1445, 2017.
51. X. Wang, Y. Yan, and R. Zhang, *Trends in Biotechnology*, Vol. 25, p. 505, 2007.
52. Y. Yan, X. Wang, Y. Pan, H. Liu, J. Cheng, Z. Xiong, F. Lin, R. Wu, R. Zhang, and Q. Lu, *Biomaterials*, Vol. 26, p. 5864 , 2005.
53. Y. Yan, X. Wang, Z. Xiong, H. Liu, F. Liu, F. Lin, R. Wu, R. Zhang, and Q. Lu, *Journal of Bioactive and Compatible Polymers*, Vol. 20, p. 259, 2005.
54. J.A. Phillippi, E. Miller, L. Weiss, J. Huard, A. Waggoner, and P. Campbell, *Stem Cells*, Vol. 26, p. 127, 2007.
55. J.D. Kim, J.S. Choi, B.S. Kim, Y.C. Choi, and Y.W. Cho, *Polymer*, Vol. 51, p. 2147, 2010.
56. S. Li, Z. Xiong, X. Wang, Y. Yan, H. Liu, and R. Zhang, *Journal of Bioactive and Compatible Polymers*, Vol. 24, p. 249, 2009.
57. S. Li, Y. Yan, Z. Xiong, C.W.R. Zhang, and X. Wang, *Journal of Bioactive and Compatible Polymers*, Vol. 24, p. 84, 2009.
58. X. Wang, K. He, and W. Zhang, *Journal of Bioactive and Compatible Polymers*, Vol. 28, p. 303, 2013.
59. Y. Huang, K. He, and X. Wang, *Materials Science and Engineering: C*, Vol. 33, p. 3220, 2013.
60. A. Nag, *Cytobios*, Vol. 28, p. 41, 1980.
61. G. Walther, J. Gekas, and O.F. Bertrand, *Catheterization and Cardiovascular Interventions*, Vol. 73, p. 917, 2009.
62. R.J. Tomanek and R.B. Runyan, *Formation of the Heart and its Regulation*, Springer Science & Business Media, 2012.

63. M. Francoise, N. Adrian, K. Ioan, and F. Gabor, *Birth Defects Research Part C: Embryo Today: Reviews*, Vol. 81, p. 320, 2008.
64. R. Gaebel, N. Ma, J. Liu, J. Guan, L. Koch, C. Klopsch, M. Gruene, A. Toelk, W. Wang, P. Mark, F. Wang, B. Chichkov, W. Li, and G. Steinhoff, *Biomaterials*, Vol. 32, p. 9218, 2011.
65. R. Gaetani, P.A. Doevendans, C.H. Metz, J. Alblas, E. Messina, A. Giacomello, and J.P. Sluijter, *Biomaterials*, Vol. 33, p. 1782, 2012.
66. B. Duan, L.A. Hockaday, K.H. Kang, and J.T. Butcher, *Journal of Biomedical Materials Research Part A*, Vol. 101A, p. 1255, 2013.
67. B. Duan, E. Kapetanovic, L.A. Hockaday, and J.T. Butcher, *Acta Biomaterialia*, Vol. 10, p. 1836, 2014.
68. T.J. Hinton, Q. Jallerat, R.N. Palchesko, J.H. Park, M.S. Grodzicki, H.-J. Shue, M.H. Ramadan, A.R. Hudson, and A.W. Feinberg, *Science Advances*, Vol. 1, 2015.
69. J. Visser, F.P.W. Melchels, J.E. Jeon, E.M. van Bussel, L.S. Kimpton, H.M. Byrne, W.J.A. Dhert, P.D. Dalton, D.W. Hutmacher, and J. Malda, *Nature Communications*, Vol. 6, p. 6933, April 2015.
70. C.H. Lee, S.A. Rodeo, L.A. Fortier, C. Lu, C. Erisken, and J.J. Mao, *Science Translational Medicine*, Vol. 6, p. 266ra171, 2014.
71. K. Joydip, S. Jin-Hyung, J. Jinah, K. Sung-Won, and C. Dong-Woo, *Journal of Tissue Engineering and Regenerative Medicine*, Vol. 9, p. 1286, 2013.
72. K.-C. Hung, C.-S. Tseng, L.-G. Dai, and S.-H. Hsu, *Biomaterials*, Vol. 83, p. 156, 2016.
73. J.H. Park, J.M. Hong, Y.M. Ju, J.W. Jung, H.-W. Kang, S.J. Lee, J.J. Yoo, S.W. Kim, S.H. Kim, and D.-W. Cho, *Biomaterials*, Vol. 62, p. 106, 2015.
74. J. Xiongfa, Z. Hao, Z. Liming, and X. Jun, *Regenerative Medicine*, Vol. 13, p. 73, 2018.
75. M.S. Mannoor, Z. Jiang, T. James, Y.L. Kong, K.A. Malatesta, W.O. Soboyejo, N. Verma, D.H. Gracias, and M.C. McAlpine, *Nano Lett.*, Vol. 13, p. 2634, June 2013.
76. H.C. Mak, *Nature Biotechnology*, Vol. 29, p. 45, January 2011.
77. W.L. Ng, S. Wang, W.Y. Yeong, and M.W. Naing, *Trends in Biotechnology*, Vol. 34, p. 689, 2016.
78. V. Lee, G. Singh, J.P. Trasatti, C. Bjornsson, X. Xu, T.N. Tran, S.-S. Yoo, G. Dai, and P. Karande, *Tissue Engineering Part C: Methods*, Vol. 20, p. 473, 2014.
79. B.R. Ringeisen, C.M. Othon, J.A. Barron, D. Young, and B.J. Spargo, *Biotechnology Journal*, Vol. 1, p. 930, 2006.
80. R.E. Saunders, J.E. Gough, and B. Derby, *Biomaterials*, Vol. 29, p. 193, 2008.
81. W. Lee, J.C. Debasitis, V.K. Lee, J.-H. Lee, K. Fischer, K. Edminster, J.-K. Park, and S.-S. Yoo, *Biomaterials*, Vol. 30, p. 1587, 2009.

82. S. Michael, H. Sorg, C.-T. Peck, L. Koch, A. Deiwick, B. Chichkov, P.M. Vogt, and K. Reimers, *PloS One*, Vol. 8, p. e57741, 2013.
83. C.P. Lu, L. Polak, A.S. Rocha, H.A. Pasolli, S.-C. Chen, N. Sharma, C. Blanpain, and E. Fuchs, *Cell*, Vol. 150, p. 136, 2012.
84. N. Liu, S. Huang, B. Yao, J. Xie, X. Wu, and X. Fu, *Scientific Reports*, Vol. 6, p. 34410, October 2016.
85. C.X.F. Lam, X.M. Mo, S.H. Teoh, and D.W. Hutmacher, *Materials Science and Engineering: C*, Vol. 20, p. 49, 2002.
86. Q. Fu, E. Saiz, and A.P. Tomsia, *Acta Biomaterialia*, Vol. 7, p. 3547, October 2011.
87. B. Guillotin, A. Souquet, S. Catros, M. Duocastella, B. Pippenger, S. Bellance, R. Bareille, M. Rémy, L. Bordenave, and J. Amédée, *Biomaterials*, Vol. 31, p. 7250, October 2010.
88. T.F. Pereira, M.A.C. Silva, M.F. Oliveira, I.A. Maia, J.V.L. Silva, M.F. Costa, and R.M.S.M. Thiré, *Virtual and Physical Prototyping*, Vol. 7, p. 275, December 2012.
89. W. Bian, D. Li, Q. Lian, W. Zhang, L. Zhu, X. Li, and Z. Jin, *Biofabrication*, Vol. 3, p. 034103, July 2011.
90. C.X.F. Lam, M.M. Savalani, S.-H. Teoh, and D.W. Hutmacher, *Biomed. Mater.*, Vol. 3, p. 034108, August 2008.
91. C.X. Lam, D.W. Hutmacher, J.T. Schantz, M.A. Woodruff, and S.H. Teoh, *Journal of Biomedical Materials Research: Part A*, Vol. 90, p. 906, 2009.
92. J. Russias, E. Saiz, S. Deville, K. Gryn, G. Liu, R. Nalla, and A. Tomsia, *J. Biomed. Mater. Res.*, Vol. 83A, p. 434, November 2007.
93. T.-S. Jang, H.-D. Jung, H.M. Pan, W.T. Han, S. Chen, and J. Song, *International Journal of Bioprinting*, Vol. 4, 2018.
94. N.E. Fedorovich, H.M. Wijnberg, W.J. Dhert, and J. Alblas, *Tissue Engineering Part A*, Vol. 17, p. 2113, 2011.
95. A. Agrawal, N. Rahbar, and P.D. Calvert, *Acta Biomaterialia*, Vol. 9, p. 5313, 2013.
96. A. Sydney Gladman, E.A. Matsumoto, R.G. Nuzzo, L. Mahadevan, and J.A. Lewis, *Nature Materials*, Vol. 15, p. 413, January 2016.
97. L.K. Narayanan, P. Huebner, M.B. Fisher, J.T. Spang, B. Starly, and R.A. Shirwaiker, *ACS Biomaterials Science & Engineering*, Vol. 2, p. 1732, 2016.
98. N. Nagarajan, A. Dupret-Bories, E. Karabulut, P. Zorlutuna, and N.E. Vrana, *Biotechnology Advances*, Vol. 36, p. 521, 2018.
99. P. Zhuang, A.X. Sun, J. An, C.K. Chua, and S.Y. Chew, *Biomaterials*, Vol. 154, p. 113 , 2018.
100. T. Guo, J.P. Ringel, C.G. Lim, L.G. Bracaglia, M. Noshin, H.B. Baker, D.A. Powell, and J.P. Fisher, *Journal of Biomedical Materials Research Part A*, Vol. 106, p. 2190, 2018.

101. T. Guo, T.R. Holzberg, C.G. Lim, F. Gao, A. Gargava, J.E. Trachtenberg, A.G. Mikos, and J.P. Fisher, *Biofabrication*, Vol. 9, p. 024101, 2017.
102. M.O. Aydogdu, N. Ekren, O. Kilic, F.N. Oktar, and O. Gunduz, 3D liquid bioprinting of the PCL/β-TCP scaffolds, in *Frontiers of Composite Materials II*, Vol. 923 of *Materials Science Forum*, pp. 79–83. Trans Tech Publications, 6 2018.
103. R.J. Jackson, P.S. Patrick, K. Page, M.J. Powell, M.F. Lythgoe, M.A. Miodownik, I.P. Parkin, C.J. Carmalt, T.L. Kalber, and J.C. Bear, *ACS Omega*, Vol. 3, p. 4342, 2018.
104. P.S. Gungor-Ozkerim, I. Inci, Y.S. Zhang, A. Khademhosseini, and M.R. Dokmeci, *Biomater. Sci.*, Vol. 6, p. 915, 2018.
105. D. Williams, P. Thayer, H. Martinez, E. Gatenholm, and A. Khademhosseini, *Bioprinting*, Vol. 9, p. 19, 2018.
106. M.M. Stanton, J. Samitier, and S. Sanchez, *Lab Chip*, Vol. 15, p. 3111, 2015.
107. E. Fennema, N. Rivron, J. Rouwkema, C. van Blitterswijk, and J. de Boer, *Trends in Biotechnology*, Vol. 31, p. 108, 2013.
108. L. Casares, R. Vincent, D. Zalvidea, N. Campillo, D. Navajas, M. Arroyo, and X. Trepat, *Nature Materials*, Vol. 14, p. 343, February 2015.
109. M.M. Stanton, J.M. Rankenberg, B.-W. Park, W.G. McGimpsey, C. Malcuit, and C.R. Lambert, *Macromolecular Bioscience*, Vol. 14, p. 953, 2014.
110. W. Xi, C.K. Schmidt, S. Sanchez, D.H. Gracias, R.E. Carazo-Salas, S.P. Jackson, and O.G. Schmidt, *Nano Letters*, Vol. 14, p. 4197, 2014.
111. S.V. Murphy and A. Atala, *Nature Biotechnology*, Vol. 32, p. 773, August 2014.
112. S. Wang, J.M. Lee, and W.Y. Yeong, *International Journal of Bioprinting*, Vol. 1, p. 3, 2015.
113. M. Wang, J. He, Y. Liu, M. Li, D. Li, and Z. Jin, *International Journal of Bioprinting*, Vol. 1, p. 15, 2015.
114. S.V. Murphy and A. Atala, *Nature Biotechnology*, Vol. 32, p. 773, August 2014.
115. S. Ahn, H. Lee, E.J. Lee, and G. Kim, *J. Mater. Chem. B*, Vol. 2, p. 2773, 2014.
116. I.T. Ozbolat and Y. Yu, *IEEE Transactions on Biomedical Engineering*, Vol. 60, p. 691, March 2013.
117. N. Hong, G.-H. Yang, J.H. Lee, and G.H. Kim, *Journal of Biomedical Materials Research Part B: Applied Biomaterials*, Vol. 106, p. 444, 2018.
118. F. Pati, J. Jang, D.-H. Ha, S. Won Kim, J.-W. Rhie, J.-H. Shim, D.-H. Kim, and D.-W. Cho, *Nature Communications*, Vol. 5, p. 3935, June 2014.

119. A. Skardal, M. Devarasetty, H.-W. Kang, I. Mead, C. Bishop, T. Shupe, S.J. Lee, J. Jackson, J. Yoo, S. Soker, and A. Atala, *Acta Biomaterialia*, Vol. 25, p. 24, 2015.
120. A. Rees, L.C. Powell, G. Chinga-Carrasco, D.T. Gethin, K. Syverud, K.E. Hill, and D.W. Thomas, *BioMed Research International*, Vol. 2015, 2015.
121. Y. Yu and I.T. Ozbolat, Tissue strands as "bioink"for scale-up organ printing, in *Engineering in Medicine and Biology Society (EMBC), 2014 36th Annual International Conference of the IEEE*, pp. 1428–1431. IEEE, 2014.
122. Z. Wu, X. Su, Y. Xu, B. Kong, W. Sun, and S. Mi, *Scientific Reports*, Vol. 6, p. 24474, April 2016.
123. A.C. Daly, G.M. Cunniffe, B.N. Sathy, O. Jeon, E. Alsberg, and D.J. Kelly, *Advanced Healthcare Materials*, Vol. 5, p. 2353, 2016.
124. M. Kesti, C. Eberhardt, G. Pagliccia, D. Kenkel, D. Grande, A. Boss, and M. Zenobi-Wong, *Advanced Functional Materials*, Vol. 25, p. 7406, 2015.
125. D.-Y. Lee, H. Lee, Y. Kim, S.Y. Yoo, W.-J. Chung, and G. Kim, *Acta Biomaterialia*, Vol. 29, p. 112, 2016.
126. M. Yeo, J.-S. Lee, W. Chun, and G.H. Kim, *Biomacromolecules*, Vol. 17, p. 1365, 2016.
127. Q. Gu, E. Tomaskovic-Crook, R. Lozano, Y. Chen, R.M. Kapsa, Q. Zhou, G.G. Wallace, and J.M. Crook, *Advanced Healthcare Materials*, Vol. 5, p. 1429, 2016.
128. M. Rimann, E. Bono, H. Annaheim, M. Bleisch, and U. Graf-Hausner, *Journal of Laboratory Automation*, Vol. 21, p. 496, 2016.
129. Y. Zhao, Y. Li, S. Mao, W. Sun, and R. Yao, *Biofabrication*, Vol. 7, p. 045002, 2015.
130. Z. Wang, R. Abdulla, B. Parker, R. Samanipour, S. Ghosh, and K. Kim, *Biofabrication*, Vol. 7, p. 045009, 2015.
131. B. Huber, K. Borchers, G.E.M. Tovar, and P.J. Kluger, *Journal of Biomaterials Applications*, Vol. 30, p. 699, 2016.
132. L. Ouyang, C.B. Highley, C.B. Rodell, W. Sun, and J.A. Burdick, *ACS Biomaterials Science & Engineering*, Vol. 2, p. 1743, 2016.
133. T.K. Merceron, M. Burt, Y.-J. Seol, H.-W. Kang, S.J. Lee, J.J. Yoo, and A. Atala, *Biofabrication*, Vol. 7, p. 035003, 2015.
134. J.W. Lee, Y.-J. Choi, W.-J. Yong, F. Pati, J.-H. Shim, K.S. Kang, I.-H. Kang, J. Park, and D.-W. Cho, *Biofabrication*, Vol. 8, p. 015007, 2016.
135. S.A. Irvine, A. Agrawal, B.H. Lee, H.Y. Chua, K.Y. Low, B.C. Lau, M. Machluf, and S. Venkatraman, *Biomedical Microdevices*, Vol. 17, p. 16, Feb 2015.
136. M. Müller, J. Becher, M. Schnabelrauch, and M. Zenobi-Wong, *Biofabrication*, Vol. 7, p. 035006, 2015.

137. S.H. Ahn, H.J. Lee, J.-S. Lee, H. Yoon, W. Chun, and G.H. Kim, *Scientific Reports*, Vol. 5, p. 13427, 2015.
138. H.J. Lee, Y.B. Kim, S.H. Ahn, J.-S. Lee, C.H. Jang, H. Yoon, W. Chun, and G.H. Kim, *Advanced Healthcare Materials*, Vol. 4, p. 1359, 2015.
139. S. Derakhshanfar, R. Mbeleck, K. Xu, X. Zhang, W. Zhong, and M. Xing, *Bioactive Materials*, Vol. 3, p. 144, 2018.
140. C.D.F. Duarte, A. Blaeser, A. Korsten, S. Neuss, J. Jäkel, M. Vogt, and H. Fischer, *Tissue Engineering Part A*, Vol. 21, p. 740, 2015.
141. F. Pati, D.-H. Ha, J. Jang, H.H. Han, J.-W. Rhie, and D.-W. Cho, *Biomaterials*, Vol. 62, p. 164, 2015.
142. C. Colosi, S.R. Shin, V. Manoharan, S. Massa, M. Costantini, A. Barbetta, M.R. Dokmeci, M. Dentini, and A. Khademhosseini, *Advanced Materials*, Vol. 28, p. 677, 2016.
143. K. Schacht, T. Jüngst, M. Schweinlin, A. Ewald, J. Groll, and T. Scheibel, *Angewandte Chemie, International Edition*, Vol. 54, p. 2816, 2015.
144. M. Kesti, M. Müller, J. Becher, M. Schnabelrauch, M. D'Este, D. Eglin, and M. Zenobi-Wong, *Acta Biomaterialia*, Vol. 11, p. 162, 2015.
145. J. Jia, D.J. Richards, S. Pollard, Y. Tan, J. Rodriguez, R.P. Visconti, T.C. Trusk, M.J. Yost, H. Yao, R.R. Markwald, and Y. Mei, *Acta Biomaterialia*, Vol. 10, p. 4323, 2014.
146. C. Xu, M. Zhang, Y. Huang, A. Ogale, J. Fu, and R.R. Markwald, *Langmuir*, Vol. 30, p. 9130, 2014.
147. E. Hoch, T. Hirth, G.E.M. Tovar, and K. Borchers, *J. Mater. Chem. B*, Vol. 1, p. 5675, 2013.
148. F.P.W. Melchels, W.J.A. Dhert, D.W. Hutmacher, and J. Malda, *J. Mater. Chem. B*, Vol. 2, p. 2282, 2014.
149. S. Catros, F. Guillemot, A. Nandakumar, S. Ziane, L. Moroni, P. Habibovic, C. van Blitterswijk, B. Rousseau, O. Chassande, J. Amédée, and J.-C. Fricain, *Tissue Engineering Part C: Methods*, Vol. 18, p. 62, 2012.
150. R. Levato, J. Visser, J.A. Planell, E. Engel, J. Malda, and M.A. Mateos-Timoneda, *Biofabrication*, Vol. 6, p. 035020, 2014.
151. S. Wüst, M.E. Godla, R. Müller, and S. Hofmann, *Acta Biomaterialia*, Vol. 10, p. 630, 2014.
152. K. Markstedt, A. Mantas, I. Tournier, H. Martínez Ávila, D. Hägg, and P. Gatenholm, *Biomacromolecules*, Vol. 16, p. 1489, 2015.
153. A.L. Rutz, K.E. Hyland, A.E. Jakus, W.R. Burghardt, and R.N. Shah, *Advanced Materials*, Vol. 27, p. 1607, 2015.
154. S. Das, F. Pati, Y.-J. Choi, G. Rijal, J.-H. Shim, S.W. Kim, A.R. Ray, D.-W. Cho, and S. Ghosh, *Acta Biomaterialia*, Vol. 11, p. 233, 2015.
155. C.R. Almeida, T. Serra, M.I. Oliveira, J.A. Planell, M.A. Barbosa, and M. Navarro, *Acta Biomaterialia*, Vol. 10, p. 613, 2014.

156. S. Seung-Joon, C. Jaesoon, P. Yong-Doo, L. Jung-Joo, H.S. Young, and S. Kyung, *Artificial Organs*, Vol. 34, p. 1044, 2010.
157. D. Loessner, C. Meinert, E. Kaemmerer, L.C. Martine, K. Yue, P.A. Levett, T.J. Klein, F.P.W. Melchels, A. Khademhosseini, and D.W. Hutmacher, *Nature Protocols*, Vol. 11, p. 727, March 2016.
158. S. Shinji, U. Kohei, G. Enkhtuul, T. Masahito, and N. Makoto, *Macromolecular Rapid Communications*, Vol. 39, p. 1700534, 2018.
159. K. Pataky, T. Braschler, A. Negro, P. Renaud, M.P. Lutolf, and J. Brugger, *Advanced Materials*, Vol. 24, p. 391, December 2011.
160. N. Celikkin, J. Simó Padial, M. Costantini, H. Hendrikse, R. Cohn, C. Wilson, A. Rowan, and W. Swieszkowski, *Polymers*, Vol. 10, p. 555, May 2018.
161. P.H.J. Kouwer, M. Koepf, V.A.A. Le Sage, M. Jaspers, A.M. van Buul, Z.H. Eksteen-Akeroyd, T. Woltinge, E. Schwartz, H.J. Kitto, R. Hoogenboom, S.J. Picken, R.J.M. Nolte, E. Mendes, and A.E. Rowan, *Nature*, Vol. 493, p. 651, January 2013.
162. A.I. Van Den Bulcke, B. Bogdanov, N. De Rooze, E.H. Schacht, M. Cornelissen, and H. Berghmans, *Biomacromolecules*, Vol. 1, p. 31, 2000.
163. M.J. Roberts, M.D. Bentley, and J.M. Harris, *Advanced Drug Delivery Reviews*, Vol. 54, p. 459, 2002. Peptide and Protein Pegylation.
164. J. Zhu, *Biomaterials*, Vol. 31, p. 4639, 2010.
165. A. Athirasala, A. Tahayeri, G. Thrivikraman, C.M. França, N. Monteiro, V. Tran, J. Ferracane, and L.E. Bertassoni, *Biofabrication*, Vol. 10, p. 024101, 2018.
166. J. Jang, H.-J. Park, S.-W. Kim, H. Kim, J.Y. Park, S.J. Na, H.J. Kim, M.N. Park, S.H. Choi, S.H. Park, S.W. Kim, S.-M. Kwon, P.-J. Kim, and D.-W. Cho, *Biomaterials*, Vol. 112, p. 264, 2017.
167. S. Das and J. Jang, *Journal of 3D Printing in Medicine*, Vol. 2, p. 69, 2018.
168. M. Izadifar, D. Chapman, P. Babyn, X. Chen, and M.E. Kelly, *Tissue Engineering Part C: Methods*, Vol. 24, p. 74, 2018.
169. C. Shikha, M. Swati, S. Aarushi, and G. Sourabh, *Advanced Healthcare Materials*, Vol. 7, p. 1701204, 2018.
170. Z. Zheng, J. Wu, M. Liu, H. Wang, C. Li, M.J. Rodriguez, G. Li, X. Wang, and D.L. Kaplan, *Advanced Healthcare Materials*, Vol. 7, p. 1701026, 2018.
171. D. Chimene, C.W. Peak, J.L. Gentry, J.K. Carrow, L.M. Cross, E. Mondragon, G.B. Cardoso, R. Kaunas, and A.K. Gaharwar, *ACS Applied Materials & Interfaces*, Vol. 10, p. 9957, 2018.
172. L. Miller, D. Chimene, G. Lokhande, and A.K. Gaharwar, Optimizing 3D printability in NICE bioinks through compositional and rheological analysis, electronic: http://loganvmiller.me/SFBPoster.pdf, 2018. Poster.

173. Wikipedia contributors, Herschel-Bulkley fluid — Wikipedia, the free encyclopedia, https://en.wikipedia.org/w/index.php?title=Herschel%E2%80%93Bulkley_fluid&oldid=807191388, 2017. [Online; accessed 24-May-2018].
174. W.H. Herschel and R. Bulkley, *Kolloid-Zeitschrift*, Vol. 39, p. 291, August 1926.
175. Y. Shi, T.L. Xing, H.B. Zhang, R.X. Yin, S.M. Yang, J. Wei, and W.J. Zhang, *Biomedical Materials*, Vol. 13, p. 035008, 2018.
176. J. Yin, M. Yan, Y. Wang, J. Fu, and H. Suo, *ACS Applied Materials & Interfaces*, Vol. 10, p. 6849, 2018.
177. D. Hu, D. Wu, L. Huang, Y. Jiao, L. Li, L. Lu, and C. Zhou, *Materials Letters*, Vol. 223, p. 219, 2018.
178. N. Faramarzi, I.K. Yazdi, M. Nabavinia, A. Gemma, A. Fanelli, A. Caizzone, L.M. Ptaszek, I. Sinha, A. Khademhosseini, J.N. Ruskin, and A. Tamayol, *Advanced Healthcare Materials*, Vol. 0, p. 1701347, 2018.
179. J.-Y. Sun, C. Keplinger, G.M. Whitesides, and Z. Suo, *Advanced Materials*, Vol. 26, p. 7608, 2014.
180. S.S. Robinson, K.W. O'Brien, H. Zhao, B.N. Peele, C.M. Larson, B.C.M. Murray, I.M.V. Meerbeek, S.N. Dunham, and R.F. Shepherd, *Extreme Mechanics Letters*, Vol. 5, p. 47, 2015.
181. K. Qiu, Z. Zhao, G. Haghiashtiani, S.-Z. Guo, M. He, R. Su, Z. Zhu, D.B. Bhuiyan, P. Murugan, F. Meng, S.H. Park, C.-C. Chu, B.M. Ogle, D.A. Saltzman, B.R. Konety, R.M. Sweet, and M.C. McAlpine, *Advanced Materials Technologies*, Vol. 3, p. 1700235, 2017.
182. J.R. Strub, E.D. Rekow, and S. Witkowski, *The Journal of the American Dental Association*, Vol. 137, p. 1289, 2006.
183. H. Sawhney and A.A. Jose, *Journal of Scientific and Technical Research*, Vol. 8, p. 1, 2018.
184. A. Tahayeri, M. Morgan, A.P. Fugolin, D. Bompolaki, A. Athirasala, C.S. Pfeifer, J.L. Ferracane, and L.E. Bertassoni, *Dental Materials*, Vol. 34, p. 192, 2018.
185. A. Dawood, B.M. Marti, V. Sauret-Jackson, and A. Darwood, *BDJ*, Vol. 219, p. 521, December 2015.
186. Q. Liu, M.C. Leu, and S.M. Schmitt, *The International Journal of Advanced Manufacturing Technology*, Vol. 29, p. 317, June 2006.
187. J.-Y. Jeng, K.-Y. Chang, D.-R. Dong, and H.S. Shih, *Rapid Prototyping Journal*, Vol. 6, p. 136, 2000.
188. J. Jeng, C. Chen, S. Chen, and R. Chen, *Journal of the Chinese Society of Mechanical Engineers, Transactions of the Chinese Institute of Engineers, Series C/Chung-Kuo Chi Hsueh Kung Ch'eng Hsuebo Pao*, Vol. 21, p. 321, 6 2000.
189. K. Torabi, E. Farjood, and S. Hamedani, *Journal of Dentistry*, Vol. 16, p. 1, May 2015.

190. C. Chen, Y. Wang, S.Y. Lockwood, and D.M. Spence, *The Analyst*, Vol. 139, p. 3219, 2014.
191. V. Mironov, T. Boland, T. Trusk, G. Forgacs, and R.R. Markwald, *Trends in Biotechnology*, Vol. 21, p. 157, 2003.
192. T. Xu, J. Jin, C. Gregory, J.J. Hickman, and T. Boland, *Biomaterials*, Vol. 26, p. 93, 2005.
193. A.-V. Do, R. Smith, T.M. Acri, S.M. Geary, and A.K. Salem, 3D printing technologies for 3D scaffold engineering in Y. Deng and J. Kuiper, eds., *Functional 3D Tissue Engineering Scaffolds*, chapter 9, pp. 203–234. Woodhead Publishing, 2018.
194. A.B. Dababneh and I.T. Ozbolat, *Journal of Manufacturing Science and Engineering*, Vol. 136, p. 061016, October 2014.
195. H. Jian, M. Wang, S. Wang, A. Wang, and S. Bai, *Bio-Design and Manufacturing*, Vol. 1, p. 45, Mar 2018.
196. Z. Xia, S. Jin, and K. Ye, *SLAS Technology: Translating Life Sciences Innovation*, Vol. 0, p. 2472630318760515, 2018.
197. B. Zhang, Y. Luo, L. Ma, L. Gao, Y. Li, Q. Xue, H. Yang, and Z. Cui, *Bio-Design and Manufacturing*, Vol. 1, p. 2, March 2018.
198. X. Cui, T. Boland, D.D. D'Lima, and M.K. Lotz, *Recent Patents on Drug Delivery & Formulation*, Vol. 6, p. 149, 2012.
199. A. Khademhosseini and G. Camci-Unal, eds., *3D Bioprinting in Regenerative Engineering: Principles and Applications*, CRC Press, Boca Raton, 2018.
200. S. Chameettachal and F. Pati, Inkjet-based 3D bioprinting in A. Khademhosseini and G. Camci-Unal, eds., *3D Bioprinting in Regenerative Engineering: Principles and Applications*, chapter 5, pp. 100–120. CRC Press, Boca Raton, 2018.
201. X. Cui, D. Dean, Z.M. Ruggeri, and T. Boland, *Biotechnology and Bioengineering*, Vol. 106, p. 963, 2010.
202. D. Poncelet, P. de Vos, N. Suter, and S.N. Jayasinghe, *Advanced Healthcare Materials*, Vol. 1, p. 27, 2012.
203. W.L. Ng, J.T.Z. Qi, W.Y. Yeong, and M.W. Naing, *Biofabrication*, Vol. 10, p. 025005, 2018.
204. Y.S. Zhang, R. Oklu, M.R. Dokmeci, and A. Khademhosseini, *Cold Spring Harbor Perspectives in Medicine*, Vol. 8, p. a025718, May 2018.
205. T. Hoshiba and J. Gong, *Microsystem Technologies*, Vol. 24, p. 613, Jan 2018.
206. P. De Coppi, G. Bartsch, Jr., M.M. Siddiqui, T. Xu, C.C. Santos, L. Perin, G. Mostoslavsky, A.C. Serre, E.Y. Snyder, J.J. Yoo, M.E. Furth, S. Soker, and A. Atala, *Nature Biotechnology*, Vol. 25, p. 100, January 2007.
207. J.-H. Shim, J.-S. Lee, J.Y. Kim, and D.-W. Cho, *Journal of Micromechanics and Microengineering*, Vol. 22, p. 085014, 2012.

208. B.-S. Kim, S.-S. Yang, and C.S. Kim, *Colloids and Surfaces B: Biointerfaces*, Vol. 170, p. 421, 2018.
209. S.-J. Lee, T. Esworthy, S. Stake, S. Miao, Y.Y. Zuo, B.T. Harris, and L.G. Zhang, *Advanced Biosystems*, Vol. 2, p. 1700213, 2018.
210. J. Ma, Y. Wang, and J. Liu, *RSC Advances*, Vol. 8, p. 21712, 2018.
211. Y. Zhang, Y. Yu, H. Chen, and I.T. Ozbolat, *Biofabrication*, Vol. 5, p. 025004, 2013.
212. F. Dolati, Y. Yu, Y. Zhang, A.M.D. Jesus, E.A. Sander, and I.T. Ozbolat, *Nanotechnology*, Vol. 25, p. 145101, 2014.
213. Q. Gao, Y. He, J.-Z. Fu, A. Liu, and L. Ma, *Biomaterials*, Vol. 61, p. 203, 2015.
214. Q. Gao, Z. Liu, Z. Lin, J. Qiu, Y. Liu, A. Liu, Y. Wang, M. Xiang, B. Chen, J. Fu, and Y. He, *ACS Biomaterials Science & Engineering*, Vol. 3, p. 399, 2017.
215. R. Attalla, C. Ling, and P. Selvaganapathy, *Biomedical Microdevices*, Vol. 18, p. 17, 2016.
216. S. Ghorbanian, M.A. Qasaimeh, M. Akbari, A. Tamayol, and D. Juncker, *Biomedical Microdevices*, Vol. 16, p. 387, 2014.
217. M. Costantini, S. Testa, P. Mozetic, A. Barbetta, C. Fuoco, E. Fornetti, F. Tamiro, S. Bernardini, J. Jaroszewicz, W. Swieszkowski, M. Trombetta, L. Castagnoli, D. Seliktar, P. Garstecki, G. Cesareni, S. Cannata, A. Rainer, and C. Gargioli, *Biomaterials*, Vol. 131, p. 98, 2017.
218. R. Chang, J. Nam, and W. Sun, *Tissue Engineering Part C: Methods*, Vol. 14, p. 157, 2008.
219. R. Chang, K. Emami, H. Wu, and W. Sun, *Biofabrication*, Vol. 2, p. 045004, 2010.
220. M. Michiya, S. Kayo, K. Koji, and A. Mitsuru, *Advanced Healthcare Materials*, Vol. 2, p. 534, 2012.
221. N.S. Bhise, V. Manoharan, S. Massa, A. Tamayol, M. Ghaderi, M. Miscuglio, Q. Lang, Y.S. Zhang, S.R. Shin, G. Calzone, N. Annabi, T.D. Shupe, C.E. Bishop, A. Atala, M.R. Dokmeci, and A. Khademhosseini, *Biofabrication*, Vol. 8, p. 014101, 2016.
222. J. Snyder, A. Rin Son, Q. Hamid, and W. Sun, *Journal of Manufacturing Science and Engineering*, Vol. 138, p. 041007, October 2015.
223. J. Zhang, F. Chen, Z. He, Y. Ma, K. Uchiyama, and J.-M. Lin, *Analyst*, Vol. 141, p. 2940, 2016.
224. W. Wu, A. DeConinck, and J.A. Lewis, *Advanced Materials*, Vol. 23, p. H178, 2011.
225. J.S. Miller, K.R. Stevens, M.T. Yang, B.M. Baker, D.-H.T. Nguyen, D.M. Cohen, E. Toro, A.A. Chen, P.A. Galie, X. Yu, R. Chaturvedi, S.N. Bhatia, and C.S. Chen, *Nature Materials*, Vol. 11, p. 768, July 2012.
226. V.K. Lee, D.Y. Kim, H. Ngo, Y. Lee, L. Seo, S.-S. Yoo, P.A. Vincent, and G. Dai, *Biomaterials*, Vol. 35, p. 8092, 2014.

227. V.K. Lee, A.M. Lanzi, H. Ngo, S.-S. Yoo, P.A. Vincent, and G. Dai, *Cellular and Molecular Bioengineering*, Vol. 7, p. 460, 2014.
228. L.E. Bertassoni, J.C. Cardoso, V. Manoharan, A.L. Cristino, N.S. Bhise, W.A. Araujo, P. Zorlutuna, N.E. Vrana, A.M. Ghaemmaghami, M.R. Dokmeci, and A. Khademhosseini, *Biofabrication*, Vol. 6, p. 024105, 2014.
229. L.E. Bertassoni, M. Cecconi, V. Manoharan, M. Nikkhah, J. Hjortnaes, A.L. Cristino, G. Barabaschi, D. Demarchi, M.R. Dokmeci, Y. Yang, and A. Khademhosseini, *Lab Chip*, Vol. 14, p. 2202, 2014.
230. S. Massa, M.A. Sakr, J. Seo, P. Bandaru, A. Arneri, S. Bersini, E. Zare-Eelanjegh, E. Jalilian, B.-H. Cha, S. Antona, A. Enrico, Y. Gao, S. Hassan, J.P. Acevedo, M.R. Dokmeci, Y.S. Zhang, A. Khademhosseini, and S.R. Shin, *Biomicrofluidics*, Vol. 11, p. 044109, 2017.
231. D.B. Kolesky, K.A. Homan, M.A. Skylar-Scott, and J.A. Lewis, *Proceedings of the National Academy of Sciences*, Vol. 113, p. 3179, 2016.
232. D.B. Kolesky, R.L. Truby, A.S. Gladman, T.A. Busbee, K.A. Homan, and J.A. Lewis, *Advanced Materials*, Vol. 26, p. 3124, 2014.
233. W. Zhu, X. Qu, J. Zhu, X. Ma, S. Patel, J. Liu, P. Wang, C.S.E. Lai, M. Gou, Y. Xu, K. Zhang, and S. Chen, *Biomaterials*, Vol. 124, p. 106, 2017.
234. X. Ma, X. Qu, W. Zhu, Y.-S. Li, S. Yuan, H. Zhang, J. Liu, P. Wang, C.S.E. Lai, F. Zanella, G.-S. Feng, F. Sheikh, S. Chien, and S. Chen, *Proceedings of the National Academy of Sciences*, Vol. 113, p. 2206, 2016.
235. C. Norotte, F.S. Marga, L.E. Niklason, and G. Forgacs, *Biomaterials*, Vol. 30, p. 5910, 2009.
236. K.A. Homan, D.B. Kolesky, M.A. Skylar-Scott, J. Herrmann, H. Obuobi, A. Moisan, and J.A. Lewis, *Scientific Reports*, Vol. 6, p. 34845, October 2016.
237. C.M. Owens, F. Marga, G. Forgacs, and C.M. Heesch, *Biofabrication*, Vol. 5, p. 045007, 2013.
238. H. Lee and D.-W. Cho, *Lab Chip*, Vol. 16, p. 2618, 2016.
239. J.U. Lind, T.A. Busbee, A.D. Valentine, F.S. Pasqualini, H. Yuan, M. Yadid, S.-J. Park, A. Kotikian, A.P. Nesmith, P.H. Campbell, J.J. Vlassak, J.A. Lewis, and K.K. Parker, *Nature Materials*, Vol. 16, p. 303, October 2016.
240. W. Liu, Z. Zhong, N. Hu, Y. Zhou, L. Maggio, A.K. Miri, A. Fragasso, X. Jin, A. Khademhosseini, and Y.S. Zhang, *Biofabrication*, Vol. 10, p. 024102, 2018.
241. P. Gatenholm, H. Backdahl, T.J. Tzavaras, R.V. Davalos, and M.B. Sano, Three-dimensional bioprinting of biosynthetic cellulose (BC) implants and scaffolds for tissue engineering, US Patent 8 691 974, assigned to Virginia Tech. Intellectual Properties, Inc. (Blacksburg, VA), April 8, 2014.
242. P. Gatenholm and D. Klemm, *MRS Bulletin*, Vol. 35, p. 208, 2010.

243. H. Bäckdahl, B. Risberg, and P. Gatenholm, *Materials Science and Engineering: C*, Vol. 31, p. 14, 2011.
244. A. Svensson, E. Nicklasson, T. Harrah, B. Panilaitis, D.L. Kaplan, M. Brittberg, and P. Gatenholm, *Biomaterials*, Vol. 26, p. 419, 2005.
245. M. Zaborowska, A. Bodin, H. Bäckdahl, J. Popp, A. Goldstein, and P. Gatenholm, *Acta Biomaterialia*, Vol. 6, p. 2540, 2010.
246. A. Bodin, S. Concaro, M. Brittberg, and P. Gatenholm, *Journal of Tissue Engineering and Regenerative Medicine*, Vol. 1, p. 406, 2007.
247. T. Mohan, T. Maver, A.D. Štiglic, K. Stana-Kleinschek, and R. Kargl, 3D bioprinting of polysaccharides and their derivatives: From characterization to application in S. Thomas, P. Balakrishnan, and M.S. Sreekala, eds., *Fundamental Biomaterials: Polymers*, Woodhead Publishing Series in Biomaterials, chapter 6, pp. 105–141. Woodhead Publishing, 2018.
248. G. Pellegrini, P. Rama, S. Matuska, A. Lambiase, S. Bonini, A. Pocobelli, R.G. Colabelli, L. Spadea, R. Fasciani, E. Balestrazzi, P. Vinciguerra, P. Rosetta, A. Tortori, M. Nardi, G. Gabbriellini, C.E. Traverso, C. Macaluso, L. Losi, A. Percesepe, B. Venturi, F. Corradini, A. Panaras, A. Di Rocco, P. Guatelli, and M. De Luca, *Regenerative Medicine*, Vol. 8, p. 553, 2013.
249. Y. Oie and K. Nishida, *Regenerative Therapy*, Vol. 5, p. 40, 2016.
250. P. Rama, S. Matuska, G. Paganoni, A. Spinelli, M. De Luca, and G. Pellegrini, *New England Journal of Medicine*, Vol. 363, p. 147, 2010.
251. G. Pellegrini, A. Lambiase, C. Macaluso, A. Pocobelli, S. Deng, G.M. Cavallini, R. Esteki, and P. Rama, *Regenerative Medicine*, Vol. 11, p. 407, 2016.
252. A. Isaacson, S. Swioklo, and C.J. Connon, *Experimental Eye Research*, May 2018.
253. A. Sorkio, L. Koch, L. Koivusalo, A. Deiwick, S. Miettinen, B. Chichkov, and H. Skottman, *Biomaterials*, Vol. 171, p. 57, 2018.
254. W. Zhu, X. Ma, M. Gou, D. Mei, K. Zhang, and S. Chen, *Current Opinion in Biotechnology*, Vol. 40, p. 103, 2016.
255. W.L. Ng, M.H. Goh, W.Y. Yeong, and M.W. Naing, *Biomater. Sci.*, Vol. 6, p. 562, 2018.
256. L. Li, Y. Zhu, and J. Yang, *Materials Letters*, Vol. 210, p. 136, 2018.
257. H. Zhao, J.H. Kwak, Y. Wang, J.A. Franz, J.M. White, and J.E. Holladay, *Carbohydrate Polymers*, Vol. 67, p. 97, 2007.
258. N. Law, B. Doney, H. Glover, Y. Qin, Z.M. Aman, T.B. Sercombe, L.J. Liew, R.J. Dilley, and B.J. Doyle, *Journal of the Mechanical Behavior of Biomedical Materials*, Vol. 77, p. 389, 2018.
259. P. Bajaj, R.M. Schweller, A. Khademhosseini, J.L. West, and R. Bashir, *Annual Review of Biomedical Engineering*, Vol. 16, p. 247, 2014.

260. J. Yan, Y. Xiao, K. Hu, S. Pan, Y. Wang, B. Zheng, Y. Wei, and L. Li, Preparation of hydrogel material for 3D bioprinting, in P. Zhao, Y. Ouyang, M. Xu, L. Yang, and Y. Ren, eds., *Applied Sciences in Graphic Communication and Packaging*, pp. 935–941, Singapore, 2018. Springer Singapore.
261. C.S. Ong, P. Yesantharao, C.Y. Huang, G. Mattson, J. Boktor, T. Fukunishi, H. Zhang, and N. Hibino, *Pediatric Research*, Vol. 83, p. 223, November 2017.
262. M. Behbehani, A. Glen, C. Taylor, A. Schuhmacher, F. Claeyssens, and J. Haycock, *International Journal of Bioprinting*, Vol. 4, 2018.
263. Y. Zhou, W. Sun, Q. Ma, and L. Zhang, Method of producing personalized biomimetic drug-eluting coronary stents by 3D-printing, US Patent 9 943 627, April 17, 2018.
264. C.X.F. Lam, R. Olkowski, W. Swieszkowski, K.C. Tan, I. Gibson, and D.W. Hutmacher, *Virtual and Physical Prototyping*, Vol. 3, p. 193, 2008.
265. H.N. Chia and B.M. Wu, *Journal of Biological Engineering*, Vol. 9, p. 4, 2015.
266. T.S. Karande, J.L. Ong, and C.M. Agrawal, *Annals of Biomedical Engineering*, Vol. 32, p. 1728, December 2004.
267. S.J. Hollister, *Nature Materials*, Vol. 4, p. 518, July 2005.
268. M.M. Stevens and J.H. George, *Science*, Vol. 310, p. 1135, 2005.
269. R.L. Truby, M. Wehner, A.K. Grosskopf, D.M. Vogt, S.G.M. Uzel, R.J. Wood, and J.A. Lewis, *Advanced Materials*, Vol. 30, p. 1706383, 2018.

6

Pharmaceutical Uses

6.1 Drug Release

The 3D printing technique in pharmaceutical manufacturing has already yielded success (1). For example, Aprecia®, an FDA-approved pharmaceutical company, has launched its first approved product, which is not only unique because of a novel manufacturing process but also better than conventional compressed tablets.

3D printing is an inexpensive additive manufacturing technique that builds a 3D object by successive layering on top of each other in a 2D fashion. The layering of the object in process is controlled digitally in a computer-aided design. The 3D printing of pharmaceuticals is best suited for personalized therapy, not only in the case of doses but dosage form as well. A personalized dosage form can be designed and printed in such a way that the drugs are combined in a single pill (polypill). This not only makes the therapy and schedule convenient for the patient but also increases adherence (1).

Traditional drug products like tablets are simple, uniform, and made for a shelf-life of 2+ years. However, 3D printing can create complex products, personalized products, and products made for immediate consumption. In August 2015 a 3D printed drug product was approved by the FDA, which is indicative of a new chapter for pharmaceutical manufacturing. The progress of 3D printed drug products has been reviewed (2).

The motivations and potential applications of 3D printed pharmaceuticals, as well as a practical viewpoint on how 3D printing could be integrated into pharmaceutical applications, have been reviewed (3).

6.1.1 *Pharmaceutical 3D Printing*

Fused deposition modeling (FDM) 3D printing has a potential to change how we envision manufacturing in the pharmaceutical industry (4). A more common utilization for FDM 3D printing is to build upon existing hot melt extrusion technology where the drug is dispersed in a polymer matrix. However, reliable manufacturing of drug-containing filaments remains a challenge along with the limitation of active ingredients which can sustain the processing risks involved in the hot melt extrusion process. To circumvent this obstacle, a single-step FDM 3D printing process was developed to manufacture thin-walled drug-free capsules which can be filled with dry or liquid drug product formulations.

The drug release from these systems is governed by the combined dissolution of the FDM capsule shell and the dosage form encapsulated in these shells (4).

To prepare the shells, the 3D printer files were modified by creating discrete zones, the so-called zoning process, with individual print parameters. Capsules printed without the zoning process resulted in macroscopic print defects and holes. X-ray computed tomography, finite element analysis and mechanical testing were used to guide the zoning process and printing parameters in order to manufacture consistent and robust capsule shell geometries (4).

6.1.2 *Pharmaceutically Acceptable Amorphous Polymers*

More recently, an attempt has been made to identify pharmaceutically acceptable amorphous polymers for producing 3D printed tablets of a model drug, haloperidol, for rapid release by fused deposition modeling (5).

Filaments for 3D printing were prepared by hot melt extrusion at 150°C with 10% and 20% haloperidol using several polymers, such as Kollidon® VA64, Kollicoat® IR, Affinsiol™15 cP, and HPMCAS. The polymers were used either individually or as binary blends.

Kollidon VA64, is a water-soluble poly(vinyl pyrrolidone)-vinyl acetate copolymer. It is commonly used to prepare solid dispersion for immediate drug release (6). Therefore, Kollidon VA64 was tested as a polymeric carrier for 3D printing. However, filaments prepared using Kollidon VA64 itself and its mixture with 10% haloperidol were brittle and collapsed in the 3D printer when pushed by a drive gear (5).

Two cellulosic fibrous polymers (HPMCAS MG and Affinisol 15 cP) were mixed individually with Kollidon VA64 for the preparation of filaments. Although HPMCAS is an enteric polymer that dissolves in aqueous media due to ionization at around pH 5.4 and higher and thus provides delayed drug release (7), it was used here to determine whether its 1:1 mixture with the highly water-soluble Kollidon VA64 could possibly provide rapid drug release.

Kollicoat IR, was also studied in conjunction with Kollidon VA64. Kollicoat IR is a poly(vinyl alcohol) (PVA), poly(ethylene glycol) (PEG) graft copolymer, and since it contains partial PEG backbone and PVA, the filaments produced by itself or its mixture with Kollidon VA64 could be suitable for 3D printing.

It was observed that all filaments produced by HPMCAS MG, Affinisol 15 cP, and Kollicoat IR by themselves or their 1:1 mixtures with Kollidon VA64 exhibited optimum mechanical properties for 3D printing.

Extruded filaments of Kollidon VA64-Affinisol 15 cP mixtures were flexible and had optimum mechanical strength for 3D printing. Tablets containing 10% drug with 60% and 100% infill showed a complete drug release at pH 2 in 45 *min* and 120 *min*, respectively. Relatively high dissolution rates were also observed at pH 6.8. The 1:1 mixture of Kollidon VA64 and Affinisol 15 cP was thus identified as a suitable polymer system for 3D printing and rapid drug release (5).

6.1.3 *Paracetamol Oral Tablets*

The manufacture of immediate release high drug loading paracetamol oral tablets could be achieved using an extrusion-based 3D printer (8). Paracetamol, also known as acetaminophen or *N*-acetyl-*p*-aminophenol, is a medicine used to treat pain and fever (9).

A premixed water-based paste formulation was used.

Poly(*N*-vinyl-2-pyrrolidone) (PVP K25) was used as a binder and croscarmellose sodium (NaCCS) (Primellose®) was used as a disintegrant.

The 3D printed tablets demonstrated a very high drug loading formulation of 80%. They can be printed as an acceptable tablet.

The 3D printed tablets were evaluated for drug release using a United States Pharmacopoeia (USP) dissolution testing type I apparatus. The tablets showed a profile characteristic of the immediate release profile. The release of 80% of the drug occurred within 5 *min*. Then, more than 90% of the paracetamol was released within the first 10 *min*.

The results of the study demonstrated the capability of 3D extrusion-based printing to produce acceptable high drug loading tablets from approved materials that comply with the current USP standards (8).

In another study, 3D printed tablets with cylindrical, gyroid lattice and bilayer structures were fabricated with customizable release characteristics (10).

Paracetamol-loaded constructs from four different pharmaceutical grade polymers, including poly(ethylene oxide), Eudragit (L100-55 and RL) and ethyl cellulose, were created using selective laser sintering 3D printing. The gyroid lattice structure was able to modulate the drug release from all four investigated polymers (10).

6.1.4 *Patient-Specific Liquid Capsules*

The pharmaceutical applications of 3D printers in pharmaceutical production have demonstrated great potential as an alternative manufacturing technique for personalizing dosage forms at a peripheral level.

A method for the production of liquid capsules with the potential of modifying drug dose and release has been presented (11).

The coordinated use of FDM, 3D printing, and liquid dispensing for the fabrication of individualized dosage form on demand in a fully automated fashion has been demonstrated.

Poly(methacrylate) shells (Eudragit EPO and RL) for immediate and extended release were fabricated using FDM 3D printing and simultaneously filled using a computer-controlled liquid dispenser loaded with model drug solution (theophylline) or suspension (dipyridamole).

The use of a shell thickness of 1.6 *mm* and a concentric architecture allowed the successful containment of a liquid core, while maintaining the release properties of the 3D printed liquid capsule. The linear relationship between the theoretical and the actual volumes from the dispenser reflected its potential for accurate dosing of R^2=0.9985.

Modifying the shell thickness of an Eudragit RL capsule allowed a controlled extended drug release without the need for a change of the formulation.

Owing to its low cost and versatility, this approach can be adapted to a wide spectrum of liquid formulations, such as small and large molecule solutions. It obviates the need for compatibility with the high temperature of FDM 3D printing process (11).

6.1.5 Thermolabile Drugs

The most commonly investigated 3D printing technology for the manufacture of personalized medicines is FDM (12). However, the high temperatures used in such a process limit its wider application.

An attempt has made to print low melting and thermolabile drugs by reducing the FDM printing temperature. Two immediate release polymers, Kollidon VA64 and Kollidon 12PF were investigated as potential candidates for low-temperature FDM printing. Ramipril was used as a model low melting temperature (109°C) drug (12). Ramipril is an angiotensin-converting enzyme inhibitor. It is used to treat high blood pressure and congestive heart failure (13).

Filaments loaded with 3% drug were obtained by hot melt extrusion at 70°C and Ramipril printlets with a dose equivalent of 8.8 *mg* were printed at 90°C (12).

It could be confirmed that the drug was stable with no signs of degradation and dissolution. The drug release from the printlets reached 100% within 20 *min* to 30 *min*.

Variable temperature Raman and solid state nuclear magnetic resonance spectroscopy techniques were used to evaluate the stability of the drug over the processing temperature range. These data indicated that Ramipril did not undergo a degradation below its melting point, which is above the processing temperature range; however, it was transformed into the impurity diketopiperazine upon exposure to temperatures higher than its melting point.

The use of the excipients Kollidon VA64 and Kollidon 12PF in FDM could be further validated by printing with the drug 4-aminosalicylic acid, cf. Figure 6.1 (12).

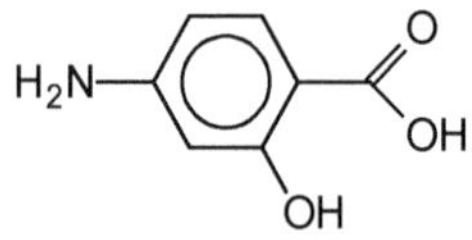

Figure 6.1 4-Aminosalicylic acid.

6.1.6 *Composite Tablets*

Composite tablets have been fabricated consisting of two components, a drug and a filler, by using a fused deposition modeling-type 3D printer (14). A PVA polymer containing calcein as model drug was used as the drug component and PVA or poly(lactic acid) (PLA) without drug was used as the water-soluble or water-insoluble filler, respectively. Calcein is also known as bis[*N*,*N*-bis(carboxymethyl)-aminomethyl] fluorescein, cf. Figure 6.2.

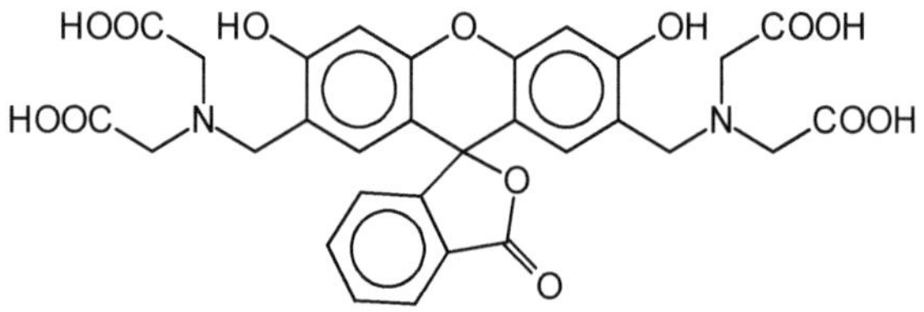

Figure 6.2 Calcein (bis[*N*,*N*-bis(carboxymethyl)-aminomethyl] fluorescein).

Various kinds of drug-PVA/PVA and drug-PVA/PLA composite tablets were designed, and the 3D printed tablets exhibited a good formability (14). The surface area of the exposed drug component is highly correlated with the initial drug release rate.

Composite tablets with an exposed top and a bottom covered with a PLA layer were fabricated. These tablets showed zero-order drug release by maintaining the surface area of the exposed drug component during drug dissolution. In contrast, the drug release profile varied for tablets whose exposed surface area changed. Composite tablets with different drug release lag times were prepared by changing the thickness of the PVA filler coating the drug component. These results, which used PVA and PLA filler, will provide useful information for preparing the tablets with multi-components and tailor-made tablets with defined drug release profiles using 3D printing (14).

6.1.7 *Transdermal Drug Delivery*

The role of two- and three-dimensional printing as a fabrication technology for sophisticated transdermal drug delivery systems has been reviewed (15).

The applicability of several printing technologies has been researched for the direct or indirect printing of microneedle arrays or for the modification of their surface through drug-containing coatings. The findings of such studies were presented. The range of printable materials that are currently used or potentially can be employed for 3D printing of transdermal drug delivery systems has also been reviewed.

Moreover, the expected impact and challenges of the adoption of 3D printing as a manufacturing technique for transdermal drug delivery systems have been assessed. Also, the current regulatory framework associated with 3D printed transdermal drug delivery systems has been outlined (15).

6.1.8 *Chip Platforms for Microarray 3D Bioprinting*

When developing therapeutic drugs, it is important to determine the safety of a drug and its efficacy (16). In the relatively early stages of drug development, drug safety and efficacy have often been tested outside the living organism, i.e., *in vitro*. The *in-vitro* assays currently available, however, use 2D cell monolayers or 3D cell spheroids, which do not adequately mimic how drugs act in the living organism, i.e., *in vivo*. Thus, an *in-vitro* cell/tissue model that can closely mimic the corresponding tissues *in vivo* and systematically simulate diseases is desired.

Microarray 3D bioprinting refers to dispensing very small amounts of cells along with other biological samples, such as hydrogels, growth factors, extracellular matrices, biomolecules, drugs, deoxyribonucleic acids, RNAs, viruses, bacteria, growth media, or combinations thereof, on a microwell/micropillar chip platform using a microarray spotter and then incubating the cells to create a mini-bioconstruct. This technology can potentially revolutionize tissue engineering and disease modeling for screening therapeutic drugs and studying toxicology.

Since microwell/micropillar chip platforms (also known as microarray biochips) contain arrays of up to 5,000 microwells or micropillars, this method is ideal for high-throughput testing. However, such currently available chips are not ideal for microarray 3D bioprinting due to the limited space available on the micropillar chip or limited control of individual experimental conditions in the microwell chip.

Recently, a micropillar chip and a microwell chip and methods for the study of cellular environments using micropillar and microwell chips have been described (16).

The micropillar chip may include at least one micropillar with a pillar-microwell. The microwell chip may include at least one microwell with an upper and a lower microwell. A perfusion channel chip that may be integrated with a micropillar chip is also disclosed. The perfusion channel chip may include a channel, a pillar-insertion hole, a membrane cassette, and a reservoir well (16).

A micropillar chip is shown in Figure 6.3, and a microwell chip is shown in Figure 6.4.

The micropillar chip has a chip base and at least one micropillar. The micropillar chip may contain arrays of micropillars, for example, about 90 to about 5,000 micropillars. The micropillar can have any shape depending on the needs of the test. For example, the micropillar may be cylindrical, cf. Figure 6.3, or it may be square.

The pillar-microwell may be capable of holding any volume of sample, including 1 μl – 4 μl.

The micropillar and microwell chips enable several methods for microarray 3D bioprinting. One exemplary method consists of dispensing cells into at least one pillar-microwell and incubating the cells to create a desired mini-bioconstruct.

The mini-bioconstructs may be created to mimic particular tissues such as a heart, liver, or brain. For example, human liver tissue

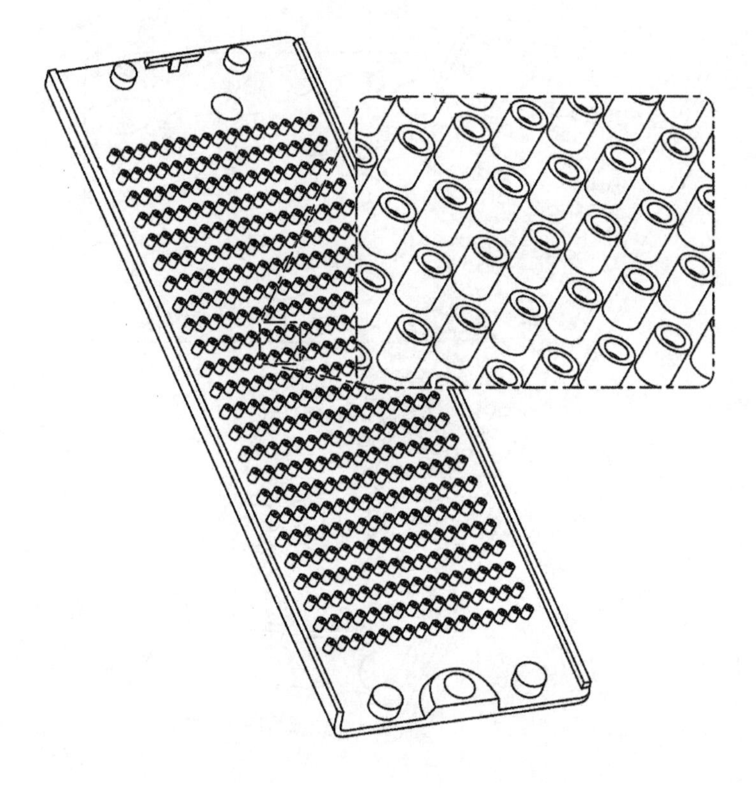

Figure 6.3 Micropillar chip (16).

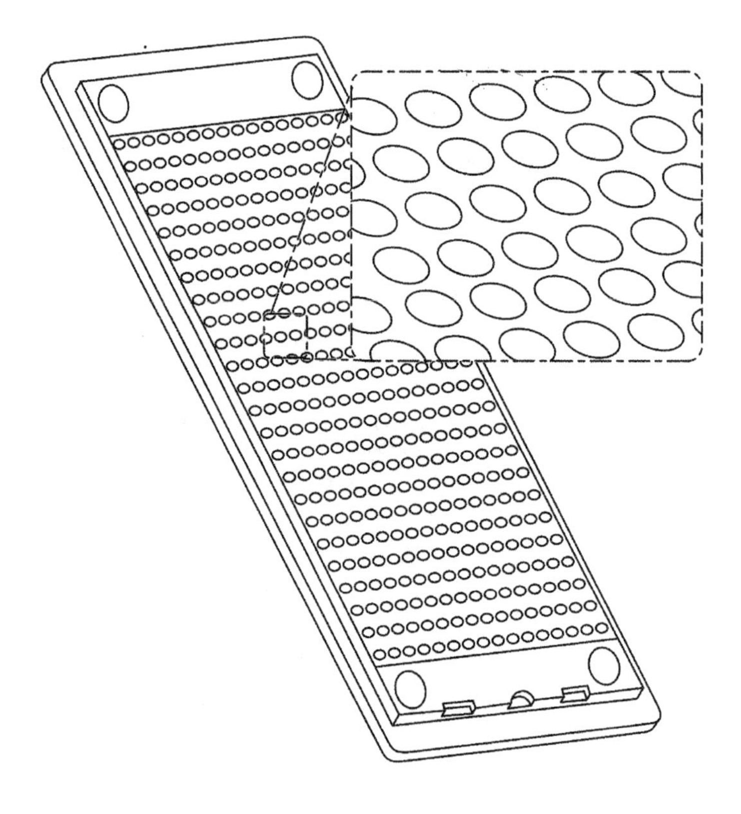

Figure 6.4 Microwell chip (16).

constructs may be created by printing primary hepatocytes/HepaRG, hepatic sinusoidal endothelial cells, hepatic stellate cells, and Kupffer cells layer-by-layer in collagen to maintain liver-specific functions. Also, human brain tissues can be generated by printing neural stem cells in Matrigel and differentiating into different neural lineages for several months.

Prior to dispensing cells, a cell suspension may be made which include the cells, at least one hydrogel, and growth media. Optionally, one or more biomolecules, drugs, DNAs, RNAs, proteins, bacteria, viruses, or combinations thereof may be included in the cell suspension. The biomolecules may be chosen to mimic a particular biological environment, such as a particular tissue.

Suitable hydrogels may be alginate, methacrylated alginate, chitosan, hyaluronic acid, fibrinogen, collagen, methacrylated collagen, PuraMatrix, Matrigel, PepGel, and PEG. The cells may be entrapped in a hydrogel using various mechanisms such as, ionic, photo, enzymatic, and chemical crosslinking. Crosslinking agents may include salts or enzymes that facilitate gelling of the hydrogel.

Examples of suitable crosslinking mechanisms and chemicals used for crosslinking are summarized in Table 6.1.

Table 6.1 Crosslinking mechanisms (16).

Crosslinking type	Chemicals
Ionic crosslinking	Alginate with barium chloride and calcium chloride
Ionic crosslinking	PuraMatrix with salts
Affinity/covalent bonding	Functionalized polymers with streptavidin and biotin
Photopolymerization	Methacrylated alginate with photoinitiators
Biocatalysis	Fibrinogen with thrombin

Among several amphiphilic polymers with maleic anhydride groups it was found that poly(maleic anhydride-*alt*-1-octadecene) provided superior coating properties with no PuraMatrix spot detachment from a micropillar chip and no air bubble entrapment in a complementary microwell chip (17). To maintain Hep3B cell viability in the PuraMatrix gel on the chip, gelation conditions were optimized in the presence of additional salts, at different seeding densities, and for growth medium washes.

As a result, salts in growth media were sufficient for gelation, and relatively high cell seeding at 6 million cells per *ml* and two media washes for pH neutralization were required (17).

An example of the preparation of mini-bioconstructs runs as follows (16):

Preparation 6–1: Mini-bioconstructs were generated by printing several layers of human cell types in photo-crosslinkable alginate with extracellular matrices and growth factors onto a 384-pillar plate containing the pillars using a microarray spotter. Hundreds of different biomimetic conditions were provided in the array of the pillars. After gelation, the 384-pillar plate was sandwiched with a 384-well plate containing growth media for rapidly testing optimum microenvironments to create human tissue replicates.

The mini-bioconstructs were then tested with compounds, stained with fluorescent dyes, and scanned with an automated fluorescent microscope for high-content imaging of organ functions and predictive assessment of the toxicity of the drug.

Also, several other examples for the preparation of similar constructs have been detailed (16).

References

1. M. Sadia, M.A. Alhnan, W. Ahmed, and M.J. Jackson, 3D printing of pharmaceuticals in W. Ahmed and M.J. Jackson, eds., *Micro and Nanomanufacturing*, Vol. 2, pp. 467–498. Springer International Publishing, Cham, 2018.
2. J. Norman, R.D. Madurawe, C.M.V. Moore, M.A. Khan, and A. Khairuzzaman, *Advanced Drug Delivery Reviews*, Vol. 108, p. 39, 2017. Editor's Collection 2016.
3. S.J. Trenfield, A. Awad, A. Goyanes, S. Gaisford, and A.W. Basit, *Trends in Pharmacological Sciences*, Vol. 39, p. 440, 2018.
4. D.M. Smith, Y. Kapoor, G.R. Klinzing, and A.T. Procopio, *International Journal of Pharmaceutics*, Vol. 544, p. 21, 2018.
5. N.G. Solanki, M. Tahsin, A.V. Shah, and A.T.M. Serajuddin, *Journal of Pharmaceutical Sciences*, Vol. 107, p. 390 , 2018.
6. P. Narayan, W. Porter, M. Brackhagen, and C. Tucker, Polymers and surfactants in A. Newman, ed., *Pharmaceutical Amorphous Solid Dispersions*, chapter 2, pp. 42–84. John Wiley & Sons, Inc, Hoboken, NJ, 2015.

7. A. Goyanes, F. Fina, A. Martorana, D. Sedough, S. Gaisford, and A.W. Basit, *International Journal of Pharmaceutics*, Vol. 527, p. 21, 2017.
8. S.A. Khaled, M.R. Alexander, R.D. Wildman, M.J. Wallace, S. Sharpe, J. Yoo, and C.J. Roberts, *International Journal of Pharmaceutics*, Vol. 538, p. 223, 2018.
9. Wikipedia contributors, Paracetamol — Wikipedia, the free encyclopedia, https://en.wikipedia.org/w/index.php?title=Paracetamol&oldid=851263704, 2018. [Online; accessed 24-July-2018].
10. F. Fina, A. Goyanes, C.M. Madla, A. Awad, S.J. Trenfield, J.M. Kuek, P. Patel, S. Gaisford, and A.W. Basit, *International Journal of Pharmaceutics*, Vol. 547, p. 44, 2018.
11. T.C. Okwuosa, C. Soares, V. Gollwitzer, R. Habashy, P. Timmins, and M.A. Alhnan, *European Journal of Pharmaceutical Sciences*, Vol. 118, p. 134, 2018.
12. G. Kollamaram, D.M. Croker, G.M. Walker, A. Goyanes, A.W. Basit, and S. Gaisford, *International Journal of Pharmaceutics*, Vol. 545, p. 144, 2018.
13. Wikipedia contributors, Ramipril — Wikipedia, the free encyclopedia, https://en.wikipedia.org/w/index.php?title=Ramipril&oldid=826948819, 2018. [Online; accessed 25-July-2018].
14. T. Tagami, N. Nagata, N. Hayashi, E. Ogawa, K. Fukushige, N. Sakai, and T. Ozeki, *International Journal of Pharmaceutics*, Vol. 543, p. 361, 2018.
15. S.N. Economidou, D.A. Lamprou, and D. Douroumis, *International Journal of Pharmaceutics*, Vol. 544, p. 415 , 2018.
16. M.-Y. Lee, Chip platforms for microarray 3D bioprinting, US Patent Application 20 180 142 195, assigned to Cleveland State University, May 24, 2018.
17. A.D. Roth, P. Lama, S. Dunn, S. Hong, and M.-Y. Lee, *Materials Science and Engineering: C*, Vol. 90, p. 634, 2018.

Index

Acronyms

Chemicals

Boldface numbers refer to Figures

General Index

Also of Interest

Check out these other books by the author published by Scrivener Publishing

Polymer Waste Management
By Johannes Karl Fink
Published 2018. ISBN 978-1-119- 53608-6

Materials, Chemicals and Methods for Dental Applications
By Johannes Karl Fink
Published 2018. ISBN 978-1-119- 51031-4

Fuel Cells, Solar Panels, and Storage Devices
Materials and Methods
By Johannes Karl Fink
Published 2018. ISBN 978-1-119-48010-5

Chemicals and Methods for Conservation and Restoration
Paintings, Textiles, Fossils, Wood, Stones, Metals, and Glass
By Johannes Karl Fink
Published 2017. ISBN 978-1-119-41824-5

Additives for High Performance Applications
Chemistry and Applications
By Johannes Karl Fink
Published 2017. ISBN 978-1-119-36361-3

Metallized and Magnetic Polymers
By Johannes Karl Fink
Published 2016. ISBN: 9781119242321

Marine, Waterborne, and Water-Resistant Polymers
Chemistry and Applications
By Johannes Karl Fink
Published 2015. ISBN 978-1-119-018486-7

The Chemistry of Printing Inks and Their Electronics and Medical Applications
By Johannes Karl Fink
Published 2015. ISBN 978-1-119-04130-6

The Chemistry of Bio-based Polymers
By Johannes Karl Fink
Published 2014. ISBN 978-1-118-83725-2

Polymeric Sensors and Actuators
By Johannes Karl Fink
Published 2012. ISBN 978-1-118-41408-8

Handbook of Engineering and Specialty Thermoplastics
Part 1: Polyolefins and Styrenics
By Johannes Karl Fink
Published 2010. ISBN 978-0-470-62483-5

Handbook of Engineering and Specialty Thermoplastics
Part 2: Water Soluble Polymers
By Johannes Karl Fink
Published 2011. ISBN 978-1-118-06275-3

A Concise Introduction to Additives for Thermoplastic Polymers
by Johannes Karl Fink.
Published 2010. ISBN 978-0-470-60955-2

CPSIA information can be obtained
at www.ICGtesting.com
Printed in the USA
BVHW041728061218
534803BV00001B/8/P